Lecture Notes in Economics and Mathematical Systems 645

For further volumes:
http://www.springer.com/series/300

Marco LiCalzi · Lucia Milone · Paolo Pellizzari
Editors

Progress in Artificial Economics

Computational and Agent-Based Models

Springer

Prof. Dr. Marco LiCalzi
Ca' Foscari University of Venice
Dept. Applied Mathematics
Cannaregio 873
30121 Venezia
Italy
licalzi@unive.it

Dipl. Oec. Lucia Milone
Ca' Foscari University of Venice
Advanced School of Economics
Cannaregio 873
30121 Venezia
Italy
lucia.milone@unive.it

Prof. Dr. Paolo Pellizzari
Ca' Foscari University of Venice
Dept. Applied Mathematics
Cannaregio 873
30121 Venezia
Italy
paolop@unive.it

ISSN 0075-8442
ISBN 978-3-642-13946-8 e-ISBN 978-3-642-13947-5
DOI 10.1007/978-3-642-13947-5
Springer Heidelberg Dordrecht London New York

Library of Congress Control Number: 2010932758

Cover design: WMXDesign GmbH, Heidelberg, Germany

Printed on acid-free paper

Springer is part of Springer Science+Business Media (www.springer.com)

Preface

This volume collects the contributions presented at the sixth event in the annual Artificial Economics conference. (For more information on former conferences, see `http://www.artificial-economics.org/`.) We received 48 submissions and based our acceptance decisions on a total of 137 referee's reports, leading to an average number of reports close to three per paper. We had to make difficult choices and we regret having been unable to include more papers.

Upon reaching a sixth anniversary, one should be able to answer a simple and basic question such as "What is Artificial Economics?" When working on this book, the editors came back to this question many times and had lively discussions leading to the conclusion that a proper answer requires a fair amount of care and ingenuity. Clearly, Artificial Economics lives in a neighborhood of Computational Economics. One of the editors likes to say that Computational Economics is an excellent first-order approximation to Artificial Economics. But what about the second order? Moreover, and more importantly, a commonly accepted definition of Computational Economics has still to come and, hence, this looks like a slippery avenue. Another editor suggests that the blend between Economics and Computer Science plays an important role in the definition of Artificial Economics scope and goals. Somehow, the *A* in Artificial rhymes well with the initials of algorithms and (economic) activities.

Artificial Economics is based on the consistent use of agent-based models and computational techniques. Virtually all contributions in this volume are interesting variations on one facet or another of these foundations. The richness and diversity underlying these models is now widely accepted as a useful companion for better understanding the experimental and theoretical results close to the heart of our scientific interests. Yet, we believe that this is not the whole story.

We would like to spell out the principle that what lies behind Artificial Economics are networks. (If you hear some background mumbling at this stage, this is a symptom of a healthy debate.) We can see at least three of these:

1. Artificial Economics connects disciplines like Economics, Management Science, and Computer Science... with a *fil rouge* emphasizing the role of agents, heterogeneity and evolution.
2. Artificial Economics links economic problems and approaches coming from different research areas, united by the need or opportunity to use simulations, numerical methods, and more generally heuristics in a broad sense;
3. Artificial Economics is made by a wide-casting network of scholars willing to recombine problems, ideas and solutions in innovative ways that draw inspiration from the areas mentioned above.

Ultimately, networks afford the multiplicity, diversity and resilience that are needed to explain our world and advance research. But the proof of the cake is in the pudding. Let us introduce the heterogenous papers appearing in this volume, conveniently (albeit somewhat arbitrarily) arranged in seven categories.

Markets and trading. Veryzhenko, Brandouy, and Mathieu tackle the question of how much sophistication is required from artificial traders to replicate well-known stylized facts in a realistic market microstructure. Hauser and Kaempff consider a market where agents are heterogeneously informed and introduce a new trading strategy that is shown to protect most of them from being exploited. Kodia, Ben Said, and Ghedira open a new front in the agent-based modeling of stylized facts for asset markets by explicitly considering behavior and cognitive attitudes.

Auctions. Brigui-Chtioui and Pinson propose a new bidding algorithm for the multicriteria English reverse auction protocol. Mochon, Saez, Gomez-Barroso, and Isasi present a simulator for the combinatorial first-price sealed-bid auction and test it over two environments inspired by current spectrum auctions. Posada and Hernández offer an agent-based perspective on recent experimental results about the performance of the continuous double auction in the presence of transaction costs.

Networks. Anand, Gai, and Marsili develop a simple model of how trust can break down in financial systems drawing on insights from the literature on coordination games and networks. Blasco and Pin study the adoption of a new technology as an instance of social learning, comparing the long-run efficiency of a network against the benchmark case of isolated agents. Taghawi-Nejad relies on a network of agents to illustrate how shocks due to the introduction of a new technology may lead to business cycles.

Management. Wall guides us into the analysis of how imperfect information affects performance under different organizational structures. Chie and Chen study different layers of the effects of social interactions on product innovation in a duopolistic dynamics. Lacagnina and Provenzano consider a multi-agent supply chain and exhibit situations of self-organized criticality that may create large fluctuations in the sector productions.

Industry Sectors. Mc Breen, Goffette-Nagot, and Jensen apply an agent-based model to provide a detailed study of the housing market that tracks the consequences of imperfect information. Schütte develops a model of product market competition and validates it using empirical data from the pharmaceutical industry in Germany. Osinga, Kramer, Hofstede, Roozmand, and Beulens investigate a complex market with many agents that is directly inspired by the Chinese pork sector.

Macroeconomics. Romanov, Yakovlev, and Lelchuk study the long-run distribution of wealth in a model with many classes of agents. Teglio, Raberto, and Cincotti report on the relationships between the availability of credit money and the variability of output and prices within the EURACE model. Hemmati, Nili, and Sadati analyze a linear-quadratic repeated inflation-unemployment game in an environment populated by heterogenous agents who use reinforcement learning to evaluate the governmental target.

Demography and culture. Giulioni and Bucciarelli apply an agent-based model to investigate the evolution of fertility and income in the process of economic development. Ruiz, Botti, Giret, Julian, Alvarado, Perez, and Rodriguez consider the effects of the labour market and of the financial sector on migration in a multi-agent simulation. Burgers, Hofstede, Jonker, and Verwaart offer a rich simulation of the impact of several cultural variables on trade.

As usual, the Conference was also enriched by two invited speakers whose (unofficial) job description is to alert us to new developments. Frank Westerhoff (University of Bamberg) gave us a wide introduction to his recent work on the use of models with heterogenous agents to probe the various effects of regulatory measures. Thomas Bäck (University of Leiden) shared with us his deep knowledge of the foundations and applications of evolutionary and bio-inspired algorithms that are becoming increasingly important for Artificial Economics and several other research areas.

To wrap things up, we would like to share that during the last hectic week when this volume was getting the final editing touches the editors were in France, Spain and Italy, respectively. Each of these three countries has played an important role in the development of the Artificial Economics series since its beginnings and, not coincidentally, this proves once again the importance of networking.

Venice,
May 2010

Marco LiCalzi
Lucia Milone
Paolo Pellizzari

Acknowledgements

Artificial Economics 2010 gratefully acknowledges the financial support of:

- AMASES (Italian Association for Mathematics in the Economic and Social Sciences);
- Banca della Marca, Orsago (Treviso);
- Department of Applied Mathematics, University Ca' Foscari Venice;
- Department of Statistics and Mathematics "Silvio Vianelli", University of Palermo;
- Italian Ministry of University and Research, PRIN Grant 2007TKLTSR;
- Fondazione Cassamarca, Treviso.

We would like to thank all the members of the Scientific Committee who generously donate their time as referees and who helped in various stages of the production of this volume:

- **Frédéric Amblard**, University of Toulouse 1 Sciences Sociales, France
- **Luis Antunes**, Universidade de Lisboa, Portugal
- **Bruno Beaufils**, University of Lille 1, France
- **Giulio Bottazzi**, Scuola Superiore Sant'Anna Pisa, Italy
- **Olivier Brandouy**, Sorbonne Graduate Business School, France.
- **Charlotte Bruun**, Aalborg Universitet, Denmark
- **Alessandro Cappellini**, Università degli Studi di Torino, Italy
- **Andrea Consiglio**, Università degli Studi di Palermo, Italy
- **Christophe Deissenberg**, Université de la Méditerraneé, France
- **Giorgio Fagiolo**, LEM, Scuola Superiore Sant'Anna, Italy
- **Josè Manuel Galàn**, Universidad de Burgos, Spain
- **Florian Hauser**, Universität Innsbruck, Austria
- **Jürgen Huber**, Universität Innsbruck, Austria
- **Wander Jager**, University of Groningen, The Netherlands
- **Marco Janssen**, Arizona State University, USA
- **Thomas Lux**, Christian-Albrechts-Universität zu Kiel, Germany
- **Rosario Mantegna**, Università degli Studi di Palermo, Italy
- **Luigi Marengo**, LEM, Scuola Superiore Sant'Anna, Italy

- **Akira Namatame**, National Defense Academy, Japan
- **Javier Pajares**, Universidad de Valladolid, Spain
- **Juan Pavòn**, Universidad Complutense de Madrid, Spain
- **Denis Phan**, GEMAS, France
- **Marta Posada**, Universidad de Valladolid, Spain
- **Juliette Rouchier**, GREQAM, France
- **Annalisa Russino**, Università degli Studi di Palermo, Italy
- **Leigh Tesfatsion**, Iowa State University, USA
- **Elpida Tzafestats**, National Technical University of Athens, Greece
- **Murat Yildizoglu**, GREQAM, France

We also thank Sonia Moulet, Paolo Pin, Oktay Sürucu, and Marco Zanella for their assistance.

Contents

List of Contributors

Oscar Alvarado
Universidad Politecnica de Valencia, Camino de Vera s/n, 46020,Valencia, Spain, e-mail: oalvarado@dsic.upv.es

Kartik Anand
The Abdus Salam International Centre for Theoretical Physics, Strada Costiera 11, 34014 Trieste, Italy, e-mail: kanand@ictp.it

Lamjed Ben Said
LI3/ENSI, Tunisia, e-mail: lamjed.bensaid@isg.rnu.tn

Adrie Beulens
Wageningen University, P.O. Box 8130, NL-6700 EW Wageningen, The Netherlands, e-mail: Adrie.Beulens@wur.nl

Andrea Blasco
Dipartimento di Scienze Economiche, Università di Bologna, Piazza Scaravilli 2, 40127 Bologna, e-mail: andrea.blasco@unibo.it

Vicente Botti
Universidad Politecnica de Valencia, Camino de Vera s/n, 46020,Valencia, Spain, e-mail: vbotti@dsic.upv.es

Olivier Brandouy
Sorbonne Graduate Business School, France, e-mail: olivier.brandouy@univ-paris1.fr

Imène Brigui-Chtioui
Institut Supérieur de Gestion, GRIISG, 147 Avenue Victor Hugo 75116 Paris, e-mail: imene.brigui-chtioui@isg.fr

Edgardo Bucciarelli
Department of Quantitative Methods and Economic Theory, University of Chieti-Pescara, Italy, e-mail: e.bucciarelli@unich.it

Saskia Burgers
Biometris Wageningen UR, Postbus 100, 6700 AC Wageningen, e-mail: saskia.burgers@wur.nl

Shu-Heng Chen
AI-ECON Research Center, Department of Economics, National Chengchi University, Taiwan, e-mail: chen.shuheng@gmail.com

Bin-Tzong Chie
Department of Industrial Economics, Tamkang University, Taiwan, e-mail: chie@mail.tku.edu.tw

Silvano Cincotti
DIBE-CINEF, Università di Genova, Via Opera Pia 11a, 16145 Genova, Italy, e-mail: cincotti@dibe.unige.it

Prasanna Gai
The Australian National University, Crawford School of Economics and Government, Canberra ACT 0200, Australia

Khaled Ghedira
LI3/ENSI, Tunisia, e-mail: khaled.ghedira@isg.rnu.tn

Adriana Giret
Universidad Politecnica de Valencia, Camino de Vera s/n, 46020,Valencia, Spain, e-mail: agiret@dsic.upv.es

Gianfranco Giulioni
Department of Quantitative Methods and Economic Theory, University of Chieti-Pescara, Italy, e-mail: g.giulioni@unich.it

Florence Goffette-Nagot
Université de Lyon, Université Lyon 2, F - 69007, Lyon, France and
CNRS, GATE Lyon-St Etienne, UMR 5824, 69130 Ecully, France, e-mail: goffette-nagot@gate.cnrs.fr

Jose Luis Gomez-Barroso
UNED and Universidad Carlos III de Madrid, Spain

Florian Hauser
Department of Banking and Finance, Innsbruck University School of Management, Universitaetsstrasse 15, A-6020 Innsbruck, Austria, e-mail: Florian.Hauser@uibk.ac.at

Mahdi Hemmati
Intelligent Systems Lab, Department of Electrical Engineering, Sharif University of Technology, Azadi Ave., Tehran, Iran, e-mail: hemmati@ee.sharif.edu

Cesareo Hernández
Escuela de Ingenierías Industriales, Universidad de Valladolid, Paseo del Cauce 59, 47011 Valladolid, Spain, e-mail: cesareo@eis.uva.es

Gert Jan Hofstede
Wageningen University, P.O. Box 8130, NL-6700 EW Wageningen, The Netherlands, e-mail: Gertjan.Hofstede@wur.nl

Pedro Isasi
UNED and Universidad Carlos III de Madrid, Spain

Pablo Jensen
ENS-LYON, Laboratoire de Physique, UMR 5672, Lyon, F-69007, France and LET, UMR 5593, University of Lyon, France, e-mail: p.jensen@ens-lyon.fr

Catholijn Jonker
Delft University of Technology, Mekelweg 4, 2628 CD Delft, NL, e-mail: c.m.jonker@tudelft.nl

Vicente Julian
Universidad Politecnica de Valencia, Camino de Vera s/n, 46020,Valencia, Spain, e-mail: vinglada@dsic.upv.es

Bob Kaempff
Department of Banking and Finance, Innsbruck University School of Management, Universitaetsstrasse 15, A-6020 Innsbruck, Austria, e-mail: Bob.Kaempff@student.uibk.ac.at

Zahra Kodia
LI3/ENSI, Tunisia, e-mail: zahra.kodia@isg.rnu.tn

Mark Kramer
Wageningen University, P.O. Box 8130, NL-6700 EW Wageningen, The Netherlands, e-mail: Mark.Kramer@wur.nl

Valerio Lacagnina
Dipartimento di Scienze Statistiche e Matematiche "Silvio Vianelli", University of Palermo, Italy, e-mail: ricopa@unipa.it

Anna Lelchuk
The Plekhanov Russian Academy of Economics, Moscow, Russia, e-mail: lelc-anya@yandex.ru

Rosa M. Rodriguez
Boeing Research & Technology Europe, Cañada Real de las Merinas, 1-3 Building 4 - 4th floor 28042 Madrid, Spain, e-mail: rosamaria.rodriguez@boeing.com

Matteo Marsili
The Abdus Salam International Centre for Theoretical Physics, Strada Costiera 11, 34014 Trieste, Italy

John Mc Breen
ENS-LYON, Laboratoire de Physique, UMR 5672, Lyon, F-69007,
France and LET, UMR 5593, University of Lyon, France, e-mail: john.

mcbreen@let.ish-lyon.cnrs.fr

Philippe Mathieu
LIFL, UMR CNRS-USTL 8022, France, e-mail: philippe.mathieu@lifl.fr

Asuncion Mochon
UNED and Universidad Carlos III de Madrid, Spain, e-mail: amochon@cee.uned.es

Masoud Nili
Graduate School of Management and Economics, Sharif University of Technology, Azadi Ave., Tehran, Iran, e-mail: m.nili@sharif.edu

Sjoukje Osinga
Wageningen University, P.O. Box 8130, NL-6700 EW Wageningen, The Netherlands, e-mail: Sjoukje.Osinga@wur.nl

Victor Perez
Boeing Research & Technology Europe, Cañada Real de las Merinas, 1-3 Building 4 - 4th floor 28042 Madrid, Spain, e-mail: victor.perezvillar@boeing.com

Paolo Pin
Dipartimento di Economia Politica, Università degli Studi di Siena, Piazza San Francesco 7, 53100 Siena, e-mail: pin3@unisi.it

Suzanne Pinson
Université Paris-Dauphine, Lamsade, Place du Maréchal de Lattre de Tassigny 75775 Paris, e-mail: suzanne.pinson@dauphine.fr

Marta Posada
Escuela de Ingenierías Industriales, Universidad de Valladolid, Paseo del Cauce 59, 47011 Valladolid, Spain, e-mail: posada@eis.uva.es

Davide Provenzano
Dipartimento di Scienze Statistiche e Matematiche "Silvio Vianelli", University of Palermo, Italy, e-mail: provenzano@unipa.it

Marco Raberto
School of Science and Engineering, Reykjavik University, Menntavegur 1, 101 Reykjavik, Iceland, e-mail: raberto@ru.is

Victor Romanov
The Plekhanov Russian Academy of Economics, Moscow, Russia, e-mail: victorromanov1@gmail.com

Omid Roozmand
Wageningen University, P.O. Box 8130, NL-6700 EW Wageningen, The Netherlands, e-mail: Omid.Roozmand@wur.nl

Nancy Ruiz
Universidad Politecnica de Valencia, Camino de Vera s/n, 46020,Valencia, Spain,

e-mail: nruiz@dsic.upv.es

Nasser Sadati
Intelligent Systems Lab, Department of Electrical Engineering, Sharif University of Technology, Azadi Ave., Tehran, Iran, e-mail: sadati@sharif.edu

Yago Saez
UNED and Universidad Carlos III de Madrid, Spain

Tino Schütte
Technische Universität Dresden, Faculty of Economics, 01062 Dresden, e-mail: tino.schuette@tu-dresden.de

Davoud Taghawi-Nejad
Department of Economics "S. Cognetti de Martiis", Via Po, 53. I-10124 Torino, Italy, e-mail: Davoud@Taghawi-Nejad.de

Andrea Teglio
Departament d'Economia, Universitat Jaume I, Av. Sos Baynat, Castellón de la Plana, Spain, e-mail: teglio@eco.uji.es

Tim Verwaart
LEI Wageningen UR, Postbus 29703, 2502 LS den Haag, NL, e-mail: tim.verwaart@wur.nl

Iryna Veryzhenko
LEM, UMR CNRS 8174, France, e-mail: iryna.veryzhenko@univ-lille1.fr

Friederike Wall
Department of Controlling and Strategic Management, School of Management and Economics, University of Klagenfurt, Austria, e-mail: friederike.wall@uni-klu.ac.at

Dmitry Yakovlev
The Plekhanov Russian Academy of Economics, Moscow, Russia, e-mail: jdemon@yandex.ru

Part I
Markets and Trading

Agent's Minimal Intelligence Calibration for Realistic Market Dynamics

Iryna Veryzhenko, Olivier Brandouy, and Philippe Mathieu

Abstract This paper investigates the question of how much sophisticated in behavior and intelligence artificial traders need to be in order to replicate both qualitative and quantitative stylized facts within a realistic market microstructure. For this purpose, we introduce an agent-based simulation environment with an architecture close to the Euronext-NYSE Stock Exchange. Series of experiments with different kinds of agents' behavior and trading framework specifications were realized within this environment. The results indicate that only special calibrations provide realistic stylized facts with coherent quantitative levels. We introduce a new type of agents, called in this paper "strongly calibrated agents", with their specific environment design, that provide price dynamics in quantitative and qualitative accordance with real stock market characteristics.

1 Introduction

Agent-Based Finance, and specifically, Agent-Based Artificial Stock Markets (hereafter ABASM) is an ever-growing field that appears, in the aftermath of the recent financial crisis, as a potential source for renewed analysis concerning the stability of the whole financial system. For example, policy experiments with agent based platforms become more realistic with the increasing sophistication of these softwares, and topics like the assessment of Tobin Tax regarding financial markets liquidity and volatility [9] or the analysis of the linkage market-microstructure and price dynamics [11] can actually be undertaken. One strong argument pleading for an increasing

Iryna Veryzhenko
LEM, UMR CNRS 8174, e-mail: iryna.veryzhenko@univ-lille1.fr

Olivier Brandouy
Sorbonne Graduate Business School, e-mail: olivier.brandouy@univ-paris1.fr

Philippe Mathieu
LIFL, UMR CNRS-USTL 8022, e-mail: philippe.mathieu@lifl.fr

M. LiCalzi et al. (eds.), *Progress in Artificial Economics*, Lecture Notes in Economics and Mathematical Systems 645, DOI 10.1007/978-3-642-13947-5_1,

role of ABASM in the academics or policy-makers toolbox, is that these softwares can duplicate the main stylized facts that can be observed on real-world stock exchanges (see for a detailed review of stylized facts [5], and for a work involving ABASM in their emergence [6]). In this perspective, several research articles have argued that Zero Intelligence Traders (ZIT, "à la" [7]) are sufficient to produce such stylized facts (for instance [15]). Nevertheless, these stylized facts remain mainly *qualitatively* congruent with real-world observations and the underlying price motions emerging from ZIT interaction remain, to our opinion, rather unrealistic. If one wants to go beyond this mere qualitative approach, ABASM need some "calibration" to deliver price dynamics that *quantitatively* correspond to real financial markets motions on the one hand, and to produce more realistic price trajectories on the other hand. Such "calibration" must be done at the agents level and matched against real-world data. Furthermore, this process must also be grounded on a realistic market microstructure. Thus, the following question is at the bottom line of the present article : *"How basic artificial traders could be for realistic market simulations ?"*.

We show, using an asynchronous Agent-Based platform benchmarked against a major European stock market, that pure ZIT cannot reproduce realistic price dynamics especially when one focus on their quantitative values. We introduce an augmented intelligence specification that aim at delivering both qualitative and quantitative stylized facts and discuss previous results that supported this minimalist specification in Agent's artificial intelligence.

The article is organized as follows. In a first section, we briefly review the relevant literature and point out the main results linking agents intelligence and financial price dynamics. We then describe the agent-based platform used for the experiments run in Sect.3, and the procedure that was used to verify the ability at mimicking its benchmark real stock market. In a fourth section, we describe our empirical strategy and present the main results we obtain.

2 Literature Review

2.1 Seminal Contribution and Initial Controversy

The minimal agents intelligence calibration became controversial when the much cited Gode and Sunder [7] model faced strong criticisms from Cliff and Bruten [4]. The main concern of this latter research was to evaluate how much intelligence is actually needed in agent to achieve high-level trading performance. Cliff and Bruten calibrated their system through agent's ability to adjust their profits in order to achieve a "realistic market efficiency". Other kinds of calibrations have appeared afterwards. For example, some researchers tried to mix agents populations in their models to get stylized facts from behavior heterogeneity (see for example [1]). Later Maslov [10] improved an existing model using limit and market orders. This method has also been employed in the work of [12]. But their models reproduce only some of generic features: namely, a congruent Hurst exponent and fat tails in the return distribution. Challet and Stinchcombe [3] show that in continuous-double auction setting the model of three processes (orders, executions, cancellations) is required to produce the fat-tails and volatility clustering.

In this research we also focus attention on price dynamics itself : this point is usually ignored by authors although we believe it is an important validation instrument for market simulator success: amazingly, one can obtain stylized facts that match "at a qualitative level" real world dynamics with an underlying price series that is totally unrealistic at a first glance. Thus, for example, even if one can observe stylized facts from the return series delivered by the simulator, he can easily face a problem of (unrealistic) highly volatile price series. Therefore, computer simulations that provide realistic stable price dynamics are particularly interesting to our opinion. The Minimal Market Model (MMM) [2] is, with respect to the latter criterion, particularly promising. These authors claim that this simple model can reproduce real market features in both price and return terms. Voit [13] also programmed this model and tried to calibrate it. Simulations end up with very volatile price series. All simplifications failed at stabilizing this highly sensitive system.

One can notice that many research claim that stylized facts, like volatility clustering, positive correlation in order types or shape of the order book, are not directly driven by strategic behavior: the necessary and sufficient ingredients to generate these statistical properties could be a specific market microstructure and zero-intelligence agents. For instance, an exhaustive investigation has been done by [8, 11], who show that the choice of a protocol may have a substantial impact on the allocative effectiveness, and other criteria such as excess volume or price dispersion. Nevertheless, and to the best of our knowledge, no paper focuses on the calibration of ABASM such to obtain quantitative stylized facts and non-volatile price dynamic in line with real markets. Our target is to fill this gap and to propose simple parametric methods to calibrate agents behavior to provide realistic price and return dynamics in the ArTificial Open Market API (here-after ATOM, see `http://atom.univ-lille1.fr`). Moving from non-strategic behavior to simple intelligence elements we show that *any assumption about any kind of intelligence has an impact to stylized facts*. We first present the ATOM API, then introduce agents behavior specification within this environment.

3 ATOM and Real World Market

ATOM is a general environment for Agent-Based simulations of stock markets. It is based on an architecture close to the Euronext-NYSE Stock Exchange one. Agent-Based artificial stock markets aim at matching orders sent by virtual traders to fix quotation prices. Price formation is ruled by a negotiation system between sellers and buyers based on an asynchronous, double auction mechanism structured in an order book. Using this API, one can generate, play or replay order flows (whatever the origin of these order flows, real world or virtual agents population). One of the main advantages of ATOM consists in its modularity. This means that it can be viewed as a system where three main components interact: i) *Agents* and their behaviors, ii) *Markets* defined in terms of microstructure and iii) the *Artificial Economic World* (including an information engine and, potentially, several economic institutions such as banks, brokers, dealers...). The two first components can be used independently or together. Depending upon the researcher targets, the *Artificial Economic World* can be plugged or not in the simulations. For example, one can use the system for the evaluation of new regulation policies or market procedures, for assessing potential effects of taxes or new trading strategies in a sophisticated artificial financial environment. Thanks to its high modularity and its ability to mimic real-world environments, it can also serve as a research tool in Portfolio Management, Algorithmic Trading or Risk Management among others. From a pure technological point of view, ATOM can also be viewed as an order-flow replay engine. This means that bankers can test their algorithmic-trading strategies using historical data without modifying the existing price series or backtest the impact of their trading-agents in totally new price motions or market regimes generated by artificial traders. Several distinctive aspects of ATOM can be highlighted:

1. It can be used without any agent. One can directly send orders written in a text file (for example, a set of orders as it arrived on a given day, for a given real-world stock market) to each order-book implemented in the simulation. In this case, ATOM serves as a "replay-engine" and simulations merely rely on market microstructure. It therefore runs really fast (an entire day of trading in less than 5 seconds).
2. ATOM can use various kind of sophisticated agents with their own behaviors and intelligence. Thousands of these agents can evolve simultaneously, creating a truly heterogeneous population. Once designed, agents evolve by themselves, learning and adapting to their (financial) environment.
3. ATOM can mix human-beings and artificial traders in a single market using its network capabilities. This allows for a wide variety of configurations, from "experimental finance" classrooms with students, to competing strategies run independently. The scheduler can be set so to allow human agents to freeze the market during their decision process or not.
4. ATOM has been tested rigorously. It has the ability to replay perfectly an order flow actually sent to a given market with the same microstructure. The resulting price series (on the one hand, the "real-world" one and on the other hand, the 'artificial" one) overlap perfectly. Moreover, given a population of agents, ATOM can generate stylized facts qualitatively similar to the market it is geared at mimicking.

Simulations in ATOM are organized as "round table discussions" and are based on an *equitably random* scheduler. Within every "round table discussion", agents are randomly interrogated using a uniformly distributed order. This latter feature ensures that each of them has an equal *possibility* of expressing its intentions. Notice that the API offers a random generator that is shared by all agents. The reproducibility of experiments is therefore guaranteed.

In real life, investors do not share the same attitudes. Some will be more reactive than others, or will implement more complex strategies leading to a higher rate of activity. In ATOM having the possibility to express an intention does not necessarily imply that a new order is issued. Since agents are autonomous, they always have the possibility to decline this opportunity.

Moreover, if an agent had been allowed to send several orders when interrogated, this would have led to an equality problem similar to the one described above. To overcome this issue, agents are just allowed to send at most *one single* order to *a given order book* (*i.e.* one order at most per stock) within the same "round table discussion". However, if an agent plans to issue several orders concerning the same stock (thus, the same order book), she must act as a finite state automata.

ATOM can include human-beings in the simulation loop. A human agent is an interface allowing for human-machine interaction. Through this interface one can create and send orders. Notice that human agents do not have any artificial intelligence: they just embed human intelligence in a formalism that is accepted by the system. To allow the introduction of human in the loop, ATOM has been designed to deal with communications over the network.

4 Empirical Strategy and Results

4.1 Data Description

Our data consist in intraday prices observed in the Euronext Paris Stock Exchange. The sampling of these observations is based on intervals of about 1 minute. We use 37 stocks for January 1st 2001 - January 31th 2001 and August 1st 2002 - August 31th 2002 (in total 1628 assets' price lists). Each day has from 1000 to 5000 records for different assets, depending on the market activity. Our goal is to compare, using price and return series, the results delivered by the ATOM platform under some set of Artificial Intelligence specifications, and real data. For this reason, we first present some general elements, then how agents behavior is progressively modified moving in the direction of growing intelligence.

4.2 Calibration Elements: Agent's Behavior

This section illustrates how we create different agents behavior and how to specify a general environment within the ATOM framework. We first start from a simple model (inspired by [10]), then step by step additional constraints are introduced in order to observe the appearance of nontrivial stylized facts and more realistic price dynamics. Some initial settings, that are implemented in the *Market* component of ATOM, are detailed below.

- Buy and Sell orders arouse with equal probability.
- Each agent can submit both orders, Buy and Sell.
- There are three possible order types: limit, market and cancel. We use also a proportion between different order types, such as 80% of limit orders, 15% of market orders and 5% of cancel orders. These proportions are equal to those observed for one specific asset within one specific day. This initial calibration is geared at imposing realistic market conditions, where both market and limit orders come in different sizes and exist for various time frame.
- Transactions are realized for a single asset.
- There are two types of traders regarding the volume that they are able to set up in the orders. A first subset, in which *"Big fish"* traders send orders with a volume close to the maximum possible value (initialization parameter), and *"Small fish"* agents, respectively with a volume close to the minimum.
- Budget constraints are implemented: traders cannot make a trade that will yield a negative profit, *i.e.*, buyers cannot buy at a price higher than their buyer value (reservation price) and sellers cannot sell for a price below their seller cost.
- Parameters for setting initialization are calculated based on the real market data. These parameters are measured for each stock within each day.

We now introduce a detailed description of agents behavior, moving from uncalibrated to strongly calibrated agents, in other words the agents in growing intelligence. These agents are realized in the *Agents* component of the ATOM API.

- *Uncalibrated agents* should be considered in the above predefined framework design. They pick a log-normally distributed price $\alpha(t) \sim Log - N(P_{mean}, P_{sd})$, where P_{mean} and P_{sd} are respectively mean value and standard deviation of real market data (initialization parameters). Volume is an integer drawn normally within the predefined (as parameters) ranges.
- *Statistically calibrated agents* have ranges within which they are able to setup orders' price and volume. Let $[P_{min}, P_{max}]$ be respectively the minimum and maximum real market intraday prices, and $[V_{min}, V_{max}]$ the minimum and maximum experienced trading volume for one specific asset. If the agent is willing to send an Ask order, he/she should respect a sellers' range $[P_{A_{min}}, P_{A_{max}}]$; respectively, in case of a Bid order, the price will be within the limits $[P_{B_{min}}, P_{B_{max}}]$, where $P_{min} \leqq P_{B_{min}} \leqq P_{A_{min}} \leqq P_{B_{max}} \leqq P_{A_{max}} \leqq P_{max}$. The price is described in the following expressions:

$$\alpha(t) = P_{A_{min}} + \triangle_\alpha \qquad \beta(t) = P_{B_{min}} + \triangle_\beta$$

where $\triangle_\alpha \sim N(P_{A_{max}} - P_{A_{min}} + 1, 1)$, $\triangle_\beta \sim N(P_{B_{max}} - P_{B_{min}} + 1, 1)$, $\alpha(t)$ and $\beta(t)$ are prices sent at the moment t by the Seller and the Buyer respectively. In a similar manner, volume is normally distributed within the limits, that are defined as follow: $V_{min} = V_{A_{min}} = V_{B_{min}}$ and $V_{max} = V_{A_{max}} = V_{B_{max}}$.

- *Strongly calibrated agents* pick a price generated with two major parameters. One parameter γ reproduces series' tendency, for instance, slow decay from maximum to minimum price during one day. The other parameter δ, normally distributed in $N(0,1)$, delivers a generic price fluctuation. An example of price formation, characterized by a slow decay, can be described as follow:

$$\gamma(0) = 1 \qquad \delta(t) \sim N(0,1)$$

$$\gamma(t) = \gamma(t-1) - 0.001 \times t$$

$$P(t) = P_{min} + (P_{max} - P_{min} + 1) \times \gamma(t) \times \delta(t)$$

  More complex price dynamics require a modification of γ parameter description. Volume, as in the previous cases, is a normally distributed value within the given range.

Using this environment and behavior calibration, we show that Uncalibrated agents fail at delivering both realistic prices and quantitative stylized facts. Thus a minimum level of calibration is necessary in the system, which directly question the fact that "zero is enough". In other terms, we show that agents actually should have non-zero intelligence, in order to perform results that are qualitatively and quantitatively congruent with empirically observed motions in real stock markets.

4.3 One Single Stock Detailed Results

This section demonstrates the simulations using BNP PARIBAS intraday price series $1^{st}August$ 2002 as a benchmark. The following list sums up the initial settings in the simulations: $P_{min} = 46.25$, $P_{max} = 48$, $P_{mean} = 47.26$, $P_{sd} = 0.35$, $V_{min} = 1$, $V_{max} = 54000$, $V_{mean} = 876.05$, $V_{sd} = 2016.19$, $Number\ of fixed\ prices = 4000$. Their applications for each calibration method are described in the following items:

- *Uncalibrated agents* pick a log-normally distributed price $\alpha(t) \sim Log - N(47.26, 0.35)$ and volume $v(t) = Log - N(876.05, 2016.19)$and randomly set transaction direction (buy/sell). There are three order types, the possibility to send a limit order is 80%, market order - 15%, cancel order - 5%.
- *Statistically calibrated agents* pick a normally distributed price within the ranges: $[45,48]$ for Ask order and $[46.25,50]$ for Bid. Transaction volume is normally distributed value between 1 and 54000. These ranges correspond to reality. These agents send the same proportion of order types as the uncalibrated agents.
- *Strongly calibrated agents* pick a price generated by the two parameters evoked previously that provide a price tendency and generic fluctuations. In addition, if we need to get three extreme maximum points in the price series, parameter γ should be coded as illustrated in the Algorithm 1.

This code defines the curve dynamics and stick price from tour to tour (tick by tick). The fluctuation, typical for real market, is provided by the parameter $\delta(t) \sim$

if $t < 50$ **then**
| $\gamma = 1 - t \times 0.02$
else if $t >= 50$ **and** $t < 100$ **then**
| $\gamma = 1.2 - t \times 0.01$
else if $t >= 100$ **and** $t < 200$ **then**
| $\gamma = 0.7 - t \times 0.001$
else if $t >= 200$ **then**
| $\gamma = 1 - t \times 0.0001$

Algorithm 1: Strong calibration

$N(0,1)$. Finally, the price is defined as $P(t) = 46.25 + (48 - 46.25 + 1) \times \gamma(t) \times \delta(t)$. The volume is defined as in the statistical calibration case. Order types proportion remains the same. In these experiments we use 20 "big fishes", that set the orders with volume from 1 to 54000 units, and 100 "small fishes" agents, that set the orders with volume from 1 to 54 units. Figure 1 shows the price plot, performed by different types of agents. As one can notice, no one can confuse price series of uncalibrated agent (even with budget constrains) with real world data. We use a Wilcoxon non-parametric test to estimate the difference in the median of generated data. According to the $p - value = 2.2e - 16$ of the Wilcoxon Test hypothesis (H_0 : ATOM generated prices series are similar to real data) the Null should be rejected. Nevertheless, our target is not to exactly duplicate real price series, but to obtain a not-so-high volatile price series, with statistical characteristics close to the real ones. Table 1 presents the statistical properties of this experiment. Quantitative characteristics coming from uncalibrated agents are far from real one. As expected, prices performed by strongly calibrated agent are much more realistic.

Quantitative properties of ATOM generated price series are far from being real. We now move to consider the properties of the return series (Fig.2). According to

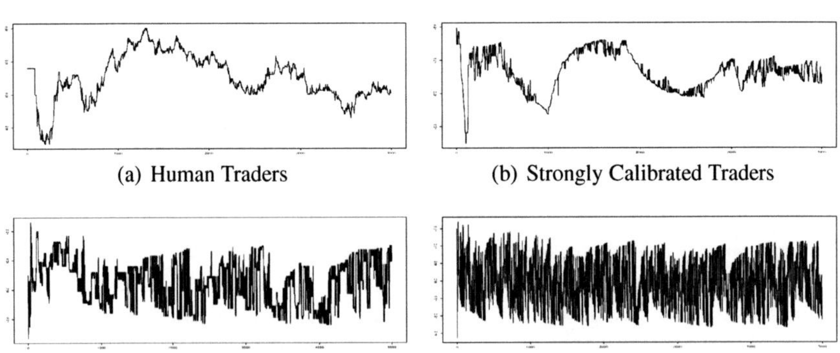

(a) Human Traders
(b) Strongly Calibrated Traders
(c) Statisticaly Calibrated Traders
(d) Uncalibrated Traders

Fig. 1 Intraday price series

Table 1 Basic statistics for price series

	Real	Strong	Statistical	Uncalibrated
Mean	47.2550	47.3195	46.2435	45.9331
Median	47.2300	47.3200	46.2600	45.9400
Variance	0.1218	0.0771	0.0198	0.4262
Stdev	0.3491	0.2777	0.1407	0.6529
Skewness	-0.2885	-0.3099	-0.0247	0.0108
Kurtosis	-0.0148	-0.0028	-0.8709	-0.9180

$p - value_{strong} = 0.9751$, $p - value_{statistical} = 0.644$, $p - value_{uncalibrated} = 0.4138$ of Wilcoxon Test, the series generated by the strongly calibrated agents are close to the real one.

Even if uncalibrated agents are able to reproduce some of the main stylized facts such as the non Gaussian return distribution (in all cases the Shapiro-Wilk test rejects the Normality hypothesis), positive autocorrelation in absolute returns (Ljung-Box test allows for the rejection of the Null "independence in a given time series"), slow decay of autocorrelation in absolute returns, the corresponding quantitative characteristics do not fit real ones (see table 2).

Uncalibrated agents' returns show high variance and standard deviation comparing with other time series, and at the same time, a very low level of kurtosis (which

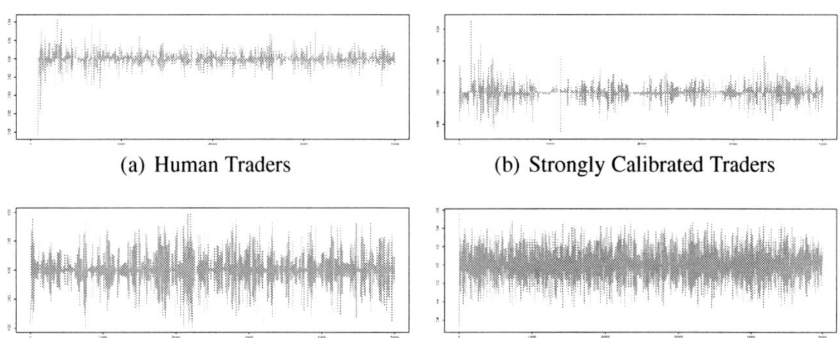

(a) Human Traders (b) Strongly Calibrated Traders

(c) Statisticaly Calibrated Traders (d) Uncalibrated Traders

Fig. 2 Intraday return

Table 2 Basic statistics of return

	Real	Strong	Statistical	Uncalibrated
Mean	-0.000002	-0.000004	-0.0000006	-0.000006
Variance	0.000000	0.000001	0.000006	0.000255
Stdev	0.000573	0.000993	0.002377	0.015963
Skewness	-0.974680	0.462673	-0.003997	0.007360
Kurtosis	21.324362	12.751463	2.987817	0.574427

is not typical for real market data), pretty close to normal one, hence high pick is not really observed in return distribution of uncalibrated agents' series. Table 2 reports the facts, that even the lack of specific strategies with simple calibration mechanisms may provide an approach of quantitative characteristics to the real ones.

4.4 Population Statistics

We now consider a population statistics or sampling distribution approach to investigate the return properties delivered by ATOM. This technique is not widely used in current financial data analysis. One example of such approach can be found in [14], who demonstrate that the sampling distribution of mean values is $Student - t$, standard deviations - χ^2 and kurtosis - Weilbull distribution.

In this research, we show that even if the statistics calculated with the ATOM generated data differ from real-world ones, once considered as population statistics, they follow the similar distributions with different parametrization (see Figs. 3, 4, 5). For this experiment, 37 assets per 22 days (a total of 814 intraday trading series) were modeled using calibration approaches described in the Sect. 4.2.

5 Conclusion

In this paper we first present our simulation environment, that allows any kind of agents behavior and artificial intelligence specification. This platform can easily be calibrated to match specific features judged as central with regards to a given real-world stock market. In this article, we use this calibration facility to investigate the following question: "What is the minimal level of artificial agents intelligence to get simulated, realistic market prices?" Based on the simulations, we show

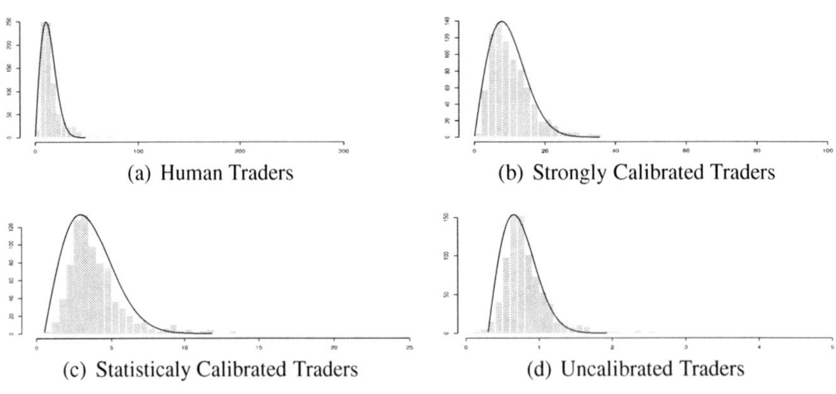

(a) Human Traders

(b) Strongly Calibrated Traders

(c) Statisticaly Calibrated Traders

(d) Uncalibrated Traders

Fig. 3 Histogram - distribution of kurtosis; Solid curve - Weilbull distribution

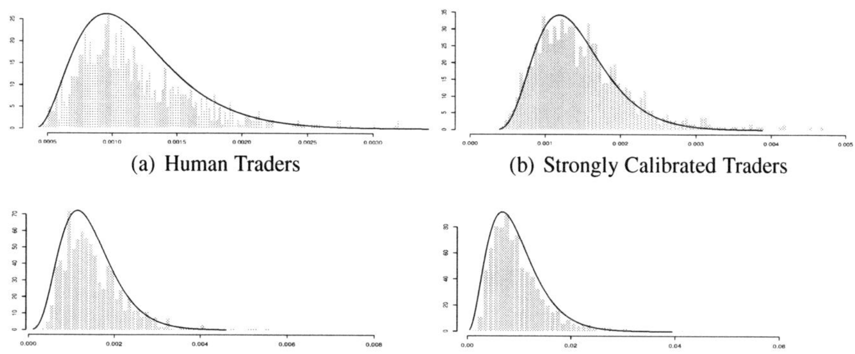

(a) Human Traders

(b) Strongly Calibrated Traders

(c) Statisticaly Calibrated Traders

(d) Uncalibrated Traders

Fig. 4 Histogram - distribution of standard deviations; Solid curve - χ^2 distribution

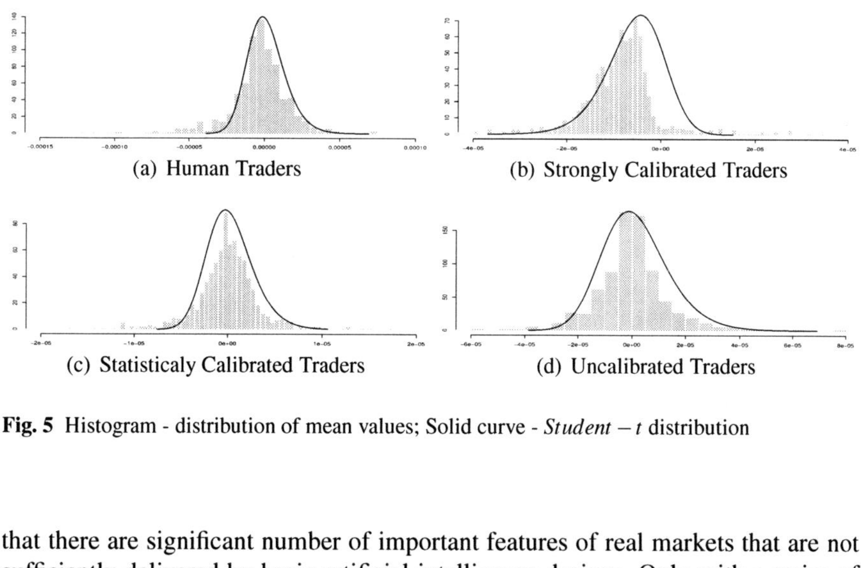

(a) Human Traders

(b) Strongly Calibrated Traders

(c) Statisticaly Calibrated Traders

(d) Uncalibrated Traders

Fig. 5 Histogram - distribution of mean values; Solid curve - $Student-t$ distribution

that there are significant number of important features of real markets that are not sufficiently delivered by basic artificial intelligence designs. Only with a series of specifications concerning agents' behavior realistic quantitative stylized facts can be obtained. Among other results, we show that *strongly calibrated agents* are definitely much less complex than human beings, and much more complex than uncalibrated (ZIT) ones. Nevertheless, results, performed by *strongly calibrated agents*, are in qualitative and quantitative agreement with empirically observed behavior of prices on real stock markets. We therefore discuss the *'zero is enough"* result that states that sophisticated behaviors are useless to understand how market motions emerge, even at the intraday level. We present an extensive empirical analysis to support this statements. From a practical point of view, this research suggests that if one wants to conduct policy-oriented experiments focusing on technical features of the market-microstructure (for example, to investigate the impact of the tick size on market liquidity and volatility), a minimal calibration of agents population is clearly necessary. Although this calibration is clearly necessary with regard to the desired properties of return series generated with agents population, a special attention should also be put on the price motions themselves.

References

1. Bak, P., M. Paczuski, Shubik, M. (1997) Price Variations in a Stock Market with Many Agents. Physica A **246**: 430–453
2. Caldarelli, C., M. Marsili, Y.-C. Zhang (1997) A Prototype Model of Stock Exchange. Europhysics Letters **40**: 479–484
3. Challet, D., Stinchcombe, R. (2001) Analyzing and modelling 1+1d markets. Physica A **300**:285
4. Cliff, D., Bruten, J. (1997) Zero is Not Enough: On The Lower Limit of Agent Intelligence For Continuous Double Auction Markets. Technical Report HPL-97-141, Hewlett-Packard Laboratories Bristol
5. Cont, R. (2001) Empirical properties of asset returns: stylized facts and statistical issues. Quantative Finance **1**: 223–236
6. Daniel, G. (2006) Asynchronous Simulations of a Limit Order Book. Ph.D. thesis, University of Manchester, UK
7. Gode, D. K., Sunder, S. (1993) Allocative Efficiency of Market with Zero-Intelligence Traders: Market as a Partial Substitute for Individual Rationality. Journal of Political Economy **101**(1): 119–137
8. LiCalzi, M., Pellizzari, P. (2006) The allocative effectiveness of market protocols under intelligent trading. Advances in Artificial Economics, pp. 17–29
9. Mannaro, K., M. Marchesi, Setzu, A. (2008) Using an artificial financial market for assessing the impact of Tobin-like transaction taxes. Journal of Economic Behavior and Organization **67**(2).
10. Maslov, S. (2000) Simple model of a limit order-driven market. Physica A **278**: 571–578
11. Pellizzari, P., Dal Forno, A. (2007) A comparison of different trading protocols in an agent-based market. Journal of Economic Interaction and Coordination **2**: 27–43
12. Preis, T., S. Golke, S. Paul, Schneider, J. (2006) Multi-agent based Order Book Model of financial markets. Europhysics Letters **75**: 510–516
13. Voit, J. (2005) The Statistical Mechanics of Financial Markets. Springer-Verlag Berlin Heidelberg
14. Watsham, T. J., Parramore, K. (1997) Quantitative methods in finance. Cengage Learning (Formerly Thomson Learning), 2nd edn
15. Yeh, C.-H. (2008) The effects of intelligence on price discovery and market efficiency. Journal of Economic Behavior and Organization **68**: 613–625

Trading on Marginal Information

Florian Hauser and Bob Kaempff

Abstract We present an agent-based simulation of a financial market with heterogeneously informed agents based on a model proposed by Schredelseker (2001). By introducing a modified fundamental trading strategy we extend the model and show that this strategy is a superior choice for most agents in the market. The modified fundamental strategy is characterized by giving more weight to the marginal piece of information an agent receives. We show that this protects agents from making joint mistakes with other market participants and suffering from a herding effect. We also observe that informational efficiency of market prices increases when agents adopt the modified trading strategy.

1 Introduction

Schredelseker [10] proposed a model where heterogeneously informed agents trade a risky asset in a call auction. He found that returns of the agents follow a non-monotonic distribution with respect to the quality of information provided to them when agents adopt a fundamental trading strategy. Lawrenz [7] showed that medium informed agents in this model realize higher losses than low-informed agents due to a herding effect. In contrast to papers like [8], herding in this model does not refer to price dynamics, but occurs when several informed agents make joint mistakes (compare [3] on this issue) by relying on fundamental information that is also known to other traders. Uninformed or relatively low-informed agents are somewhat protected from this effect as they hardly get any information which could induce a bias in their estimation. Or, as pointed out in [9], "traders with less information seem to be in advantage because of that part of information which is unknown to them".

Florian Hauser · Bob Kaempff
Department of Banking and Finance, Innsbruck University School of Management, Universitaetsstrasse 15, A-6020 Innsbruck, Austria,
e-mail: Florian.Hauser@uibk.ac.at, Bob.Kaempff@student.uibk.ac.at

M. LiCalzi et al. (eds.), *Progress in Artificial Economics*, Lecture Notes in Economics and Mathematical Systems 645, DOI 10.1007/978-3-642-13947-5_2,

The finding that herding worsens returns for medium-informed agents goes hand in hand with negative effects on informational efficiency of market prices in this model. In that sense this issue is closely linked to the Overvaluation Hypothesis [1]: if too many traders in a market react to one specific piece of information, this may lead to biased market prices. To deal with this issue, [11] proposed two alternative trading strategies that completely ignore information and showed that some agents in this model can improve their performance when adopting them. For a related model governing a large number of agents, [5] showed that ignoring fundamental information and adopting a random trading strategy is a dominant choice for most agents in the market. Both authors also find that informational efficiency of market prices improves when agents optimize their payoffs by adopting alternative trading strategies. On the other hand, several papers also report a negative marginal utility for random trading strategies. The more traders decide to ignore information, the worse returns for them will be (compare [4, 6]).

In this paper we extend the Schredelseker-model by introducing a modified fundamental strategy that avoids joint mistakes by concentrating on the marginal piece of information provided to an agent. In Sect. 2 we briefly discuss the market model. Sect. 3 presents results based on a pure fundamental trading strategy and explains the resulting herding effect. Sect. 4 introduces the modified trading strategy that avoids herding. We show that this strategy improves the performance of less-informed agents and makes market prices become more efficient. Sect. 5 concludes.

2 Market Model

Our market model closely follows the original model proposed by [10] where one risky asset is traded on a one-period call market. The fundamental value v of the asset is determined by ten binomial signals ε_i:

$$v = \sum_{i=1}^{10} \varepsilon_i;\ \varepsilon \in \{0,1\} \text{ with } \mathrm{P}(\varepsilon = 0) = \mathrm{P}(\varepsilon = 1). \qquad (1)$$

The market is populated by ten risk-neutral agents $T \in \{1,2,\dots 10\}$ who are bound to submit reservation prices CV_T for one share of the contract, indicating that they are willing to buy when the market price $P < CV_T$ or sell if $P > CV_T$. Based on all reservation prices, a market clearing price P that maximizes trading volume is fixed.[1] After trading, all shares are liquidated according to their fundamental value v, so gains or losses of one agent can be calculated as:

$$R_T = (v - P) \times s_T, \qquad (2)$$

[1] In most situations the median reservation price serves as a market clearing price. In situations where several agents submit reservation prices equal to the median reservation price, P is adjusted to $P - 0.05$ or $P + 0.05$ if one of these options generates a higher trading volume.

with s_T being 1 (-1) if agent T bought (sold) one share.[2]

All ten agents are provided with different information on the fundamental value v, with agent T knowing the realization of all signals $\varepsilon_{i \leq T}$. Hence, agent $T = 1$ only knows the realizations of the first signal ε_1 while agent $T = 10$ is completely informed as he knows the realization of all ten signals. According to this, the information system in this market is cumulative in the sense that any signal agent $T = x$ receives is also included in the information set provided to a better informed agent $T > x$.

All results provided below are based on 1024 runs r, each run refers to one unique realization of the ten signals ε_i. As the variance of returns captures squared deviations of market prices from the true value of the contract, it will serve us as a proxy for market inefficiency in the following:

$$\sigma_M^2 = \frac{\sum_{\text{run}=1}^{1024} (v - P)^2}{1024}. \tag{3}$$

The insider $T = 10$ is perfectly informed in our model, and he is able to calculate the fundamental value of the contract in each run. Considering now the definition of [2], our market is fully efficient only if market prices are identical to the fundamental value in each run and $\sigma_M^2 = 0$.

3 Original Fundamental Trading Strategy

When relying on the original fundamental trading strategy as proposed by [11], agent T calculates the expected value of the contract with respect to the signals provided to him:

$$CV_T = \sum_{i=1}^{T} \varepsilon_i + 0.5 \times (10 - T). \tag{4}$$

When all agents in the market do so, gains and losses of the agents will follow a j-shaped distribution as plotted in Fig. 1.

We already argued that the high losses realized by the medium-informed agents can be attributed to a herding effect that emerges when agents process signals that are also known to other agents. If one piece of information is processed by too many agents, this may result in an overreaction of market prices as described in [1]. Calculating market inefficiency as defined in Eq. (3) supports our argumentation of market prices being biased: the relatively high value of $\sigma_M^2 = 0.76$ corresponds to the average prediction error of an agent who knows only seven signals,[3] so market prices clearly not reflect all available information on the fundamental value of the contract traded.

---

[2] If the number of buyers and sellers differs, orders will be rationed.

[3] The prediction error follows the same logic as the formula for market inefficiency (3) and is calculated as $PE_T = \frac{\sum_{\text{run}=1}^{1024} (v - CV_T)^2}{1024}$.

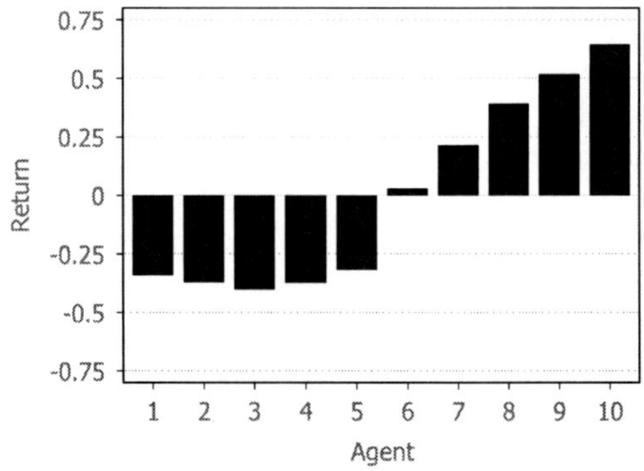

Fig. 1 Average returns for all ten agents adopting a fundamental strategy

To shed further light on the herding mechanism we introduce ten vectors $\overrightarrow{D_{T;r}}$, one for each agent, taking a value of 1 if agent T buys the contract in run r, –1 if she sells, and 0 if the agent does not trade in this run. For every agent T we calculate the correlation of his vector with the ten signal vectors $\overrightarrow{S_{i;r}}$, with $S_{i;r}$ taking the value of ε_i in run r. This leaves us with 100 correlations in total, denoted as $\rho_{T;i}$ in the following. The motivation for this is to find a proxy that indicates the impact of each signal on the trading decisions of each agent in the market. Intuitively we assume that the signal vectors known to a given agent are positively correlated with his likelihood of buying the contract. E.g., if agent $T = 2$ observes $\varepsilon_2 = 1(0)$ in a given run, he will submit a relatively high (low) reservation price. This increases the chance for him to be a buyer (seller), so $\rho_{2;2}$ should take a positive value.

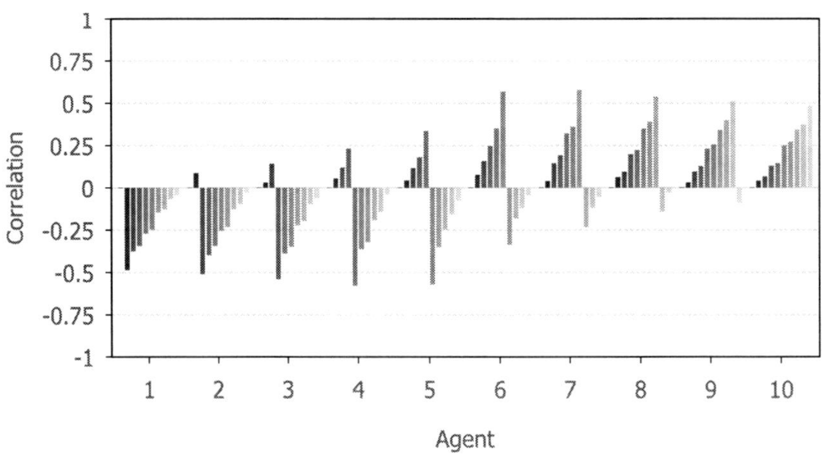

Fig. 2 Correlations $\rho_{T;i}$ for all agents when adopting a fundamental trading strategy. Black bars denote correlations with ε_1 while the lightest grey bars denote correlations with ε_{10}

All resulting 100 correlations are reported in Fig. 2. Note that $\rho_{i;1} = 0$ for every agent in the market, thus there are only 9 correlations per agent visible in Fig. 2. As signal ε_1 is processed by all agents, it can be considered as public knowledge. When $\varepsilon_1 = 1(0)$ all agents will submit a higher (lower) reservation price, hence ε_1 has no influence on any agent's trading decision.

In line with our argumentation above we observe positive correlations for all signals an agent processes (except for ε_1), showing that a positive (negative) value of one signal increases the probability for this agent to buy (sell). E.g., for agent $T = 2$ (as indicated on the horizontal axis) the first bar (dark black) refers to the positive correlation $\rho_{2;2}$ mentioned above, while the other bars associated with that agent (dark to light grey) indicate negative correlations with the signals unknown to him.

When comparing the strength of positive correlations we find them to decrease for lower-order signals. I.e. the second signal ε_2 shows only very limited influence on any agent (including also $T = 2$ who can separate himself from other agents *only* by this signal) to become buyer or seller. As this piece of information is processed by nine out of ten agents it can be considered as being almost public knowledge. Hence, the reservation prices of nine agents will be correlated as all of them process this piece of information. As a consequence, this signal will have considerable impact on market prices and herding will occur. For any given agent we observe his marginal signal ε_T to have the highest impact on his trading decision.

On the other hand we observe $D_{T;r}$ to be negatively correlated with all signals unknown to the respective agent. The negative correlations indicate that an agent tends to sell (buy) when a hidden signal points to a high (low) asset value. This results as the hidden signals are processed by better informed agents. The latter tend to make correct decisions whether to buy or to sell and their decisions will move the market price away from the estimation of the less-informed.

4 Modified Fundamental Trading Strategy

In this section we first introduce and define a modification of the fundamental trading strategy. We then show results emerging when only one of the ten agents in the market adopts this strategy. In the last subsection we present results arising when several agents in the market adopt the modified fundamental trading strategy.

4.1 Definition

The positive correlation between $D_{T;r}$ and the individual signals is always strongest for the marginal signal of the respective agent, e.g. agent $T = 4$ mainly buys when signal ε_4 points to a high asset value. As discussed above, less-informed agents trading on lower-order signals are likely to induce a herding-effect. In that case those agents will make correlated errors and the market will redistribute wealth away from them as described by [3]. This leads us to the conclusion that agents should rely more strongly on their marginal signal to enhance their returns. To transform this intuition to a trading strategy consider the formula denoted as "modified fundamental strategy" in the following:

$$CVM_T = \sum_{i=1}^{T-1} \varepsilon_i + Z \times \varepsilon_T + 0.5 \times (11 - Z - T). \tag{5}$$

The first expression $\sum_{i=1}^{T-1} \varepsilon_i$ is part of the original fundamental trading strategy as proposed by [10] and takes into account all observed signals except the marginal signal ε_T. The second expression $Z \times \varepsilon_T$ introduces a weighting coefficient $Z \in [1;10]$, giving an agent the possibility to change the weight on his marginal signal ε_T. The last part of the formula, $0.5 \times (11 - Z - T)$, accounts for the unobserved signals $\varepsilon_{i>T}$ and balances the reservation prices similarly to the original fundamental strategy. Note that with $Z = 1$, the modified fundamental trading strategy is identical to the original one. The higher Z, the more the marginal signal dominates lower-order signals when calculating the reservation price.

Consider an example for agent $T = 4$ with $Z = 4$ and $\varepsilon_{\{1,2,3\}} = \{0,1,0\}$: The difference between the original and the modified fundamental strategy depends solely on the marginal signal ε_4 that this agent receives. When $\varepsilon_4 = 1$, the fundamental trading strategy will result in a reservation price of $CV_4 = 0+1+0+1+3 = 5$, when $\varepsilon_4 = 0$ the reservation price will be $CV_4 = 4$. The mean of both possible reservation prices CV_4 is 4.5. Applying the modified fundamental trading strategy, $CVM_4 = 0+1+0+4\times 1+1.5 = 6.5$ when $\varepsilon_4 = 1$ and $CVM_4 = 2.5$ for $\varepsilon_4 = 0$. Note that the mean of both possible reservation prices CVM_4 is still 4.5, which demonstrates that the modified fundamental strategy does not bias the expected reservation price. It just increases the sensitivity of the reservation price to the marginal signal an agent receives.

4.2 Optimization for one Agent

To analyze if agents can improve their performance by applying the modified fundamental trading strategy we start with the situation where all agents rely on the original fundamental strategy as proposed by [10]. Table 1 reports returns for all agents when applying the modified fundamental strategy with different weighting coefficients Z while all other agents in the market stay with the original fundamental trading strategy. With this ceteris paribus assumption, we find that any strategy with $Z > 1$ is a dominant choice over the fundamental strategy for all agents $T < 8$.

To provide the reader with a better intuition for the impact of the modified fundamental trading strategy on the returns of an agent as well as on market efficiency, we exemplarily pick agent $T = 4$. Assuming that all other agents stay with the original fundamental strategy, Fig. 3 reports returns for $T = 4$ (black line) as well as

Table 1 Returns for any given agent that adopts the modified fundamental trading strategy while all other agents rely on the original fundamental trading strategy

T	1	2	3	4	5	6	7	8	9	10
Z=1	−0.340	−0.371	−0.401	−0.372	−0.317	0.030	0.213	0.394	0.518	0.645
Z=2	−0.317	−0.303	−0.265	−0.199	−0.037	0.155	0.252	0.360	0.471	0.580
Z=3	−0.241	−0.189	−0.125	−0.012	0.100	0.195	0.270	0.321	0.385	0.477
Z=4	−0.161	−0.105	−0.035	0.043	0.103	0.196	0.271	0.321	0.350	0.405
Z=5	−0.112	−0.060	−0.008	0.043	0.103	0.196	0.271	0.321	0.350	0.380
Z=6	−0.090	−0.047	−0.008	0.043	0.103	0.196	0.271	0.321	0.350	0.380

market inefficiency as defined in Eq. (3) (grey line) for various weighting coefficients $Z \in [1;6]$. Note that results for $Z = 1$ correspond to the situation where all ten agents adopt the original fundamental trading strategy.

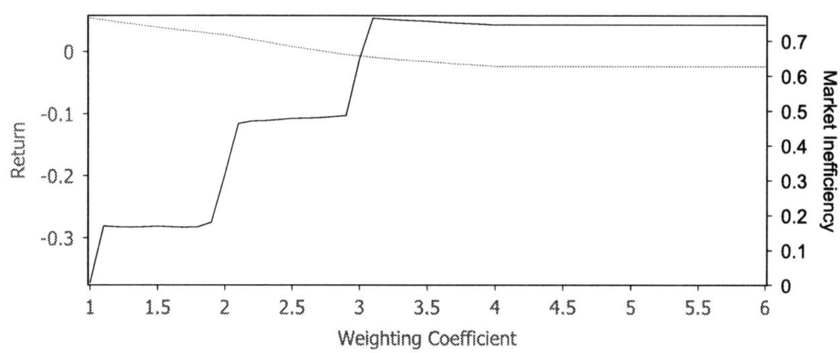

Fig. 3 Average returns for $T = 4$ adopting the modified fundamental trading strategy with different weighting coefficients Z (black line) and resulting market inefficiency (grey line). All other agents stick with the original fundamental trading strategy

We observe that the return for $T = 4$ monotonically increases up to $Z = 3.1$, which marks the optimal choice for this agent in this situation. For values of $Z > 3.1$ returns decrease again, but note that a modified fundamental trading strategy with any weighting coefficient $Z > 1$ will still leave the agent with a higher return than the original fundamental strategy.

Turning to market inefficiency we find no specific optimum. σ_M^2 monotonically decreases up to values of $Z = 4$. For all weighting factors $Z > 4$, market inefficiency remains at a level of $\sigma_M^2 = 0.63$. This result shows that market prices will become more informative when agent $T = 4$ applies the modified fundamental trading strategy instead of the original one.

To explain the success of the modified fundamental trading strategy in terms of improving returns and market efficiency, we consider again the correlations $\rho_{T;i}$ as shown above in Fig. 2. Staying with our example where *only* agent $T = 4$ adopts the modified fundamental trading strategy, Fig. 4 shows all ten correlations $\rho_{4;i}$ resulting from different weighting coefficients Z for the modified fundamental trading strategy of $T = 4$.

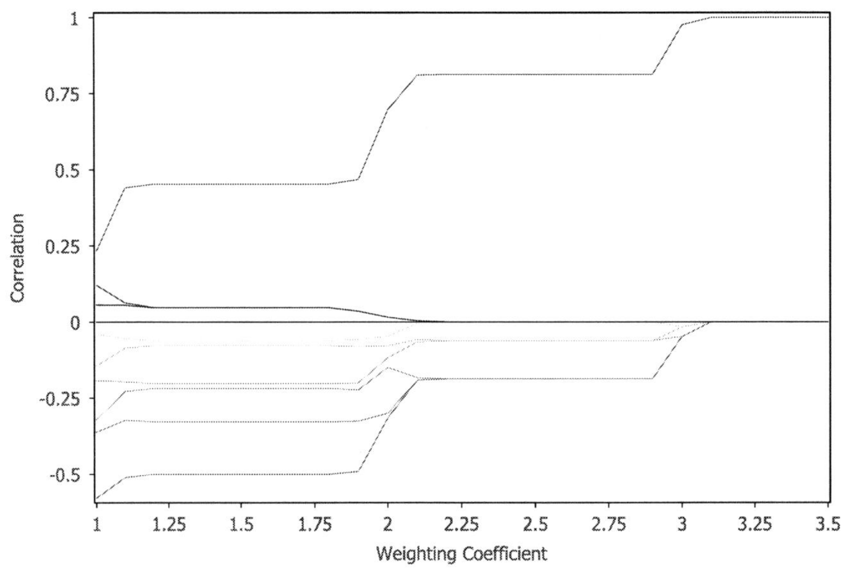

Fig. 4 Correlations $\rho_{4;i}$ emerging when agent $T = 4$ applies the modified fundamental trading strategy with different weighting coefficients Z while all other agents stick with the original fundamental trading strategy. Black lines denote correlations with ε_1 while the lightest grey lines denote correlations with ε_{10}

The values of $\rho_{4;i}$ for a weighting coefficient of $Z = 1$ correspond to the situation where also $T = 4$ adopts the original fundamental trading strategy. Adopting a modified fundamental strategy with increasing values for the weighting coefficient Z induces

- monotonically increasing values for $\rho_{4,i}; i = 4$, referring to the marginal signal agent $T = 4$ receives.
- monotonically decreasing values for $\rho_{4;i}; i < 4$, indicating that lower-order signals become less important for the trading decision of agent $T = 4$.
- increasing (negative) values for $\rho_{4,i}; i > 4$, referring to the signals not known to agent $T = 4$.

This result strongly supports our hypothesis that the modified fundamental trading strategy reduces the herding effect for agent $T = 4$: when he places more weight on his marginal piece of information, the lower-order signals that induce herding become less important. At the same time, this agent reduces the probability of being exploited by better-informed agents as indicated by the increasing negative correlations with the hidden signals.

For a weighting coefficient of $Z \geq 3.1$ we observe $\rho_{4;4} = 1$ while all other correlations are 0. This shows that now the trading decisions of this agent solely depend on the realization of his marginal signal: when it indicates a high asset value ($\varepsilon_4 = 1$), agent $T = 4$ will buy the asset, and when it points to a low value this agent will sell. Doing so he is completely immune to herding, and with $Z = 3.1$ this strategy generates the highest possible return for him.

4.3 Equilibrium for two Possible Trading Strategies

Up to this point we considered only one agent that applies the modified fundamental trading strategy. In this subsection we lift this restriction and describe a situation where several agents try to optimize their trading strategies.

As soon as one agent in the market changes his trading strategy, this will affect the optimal strategy choice for all other agents in the market. In that sense, the optimization procedure for our agents is a game-theoretic problem. A complete analysis of possible strategy equilibria would go beyond the scope of this article and must be left for future research.

At this stage we will only consider a sequential optimization of all agents that is limited to the original fundamental trading strategy as well as the modified fundamental trading strategy with a fixed weighting coefficient $Z = 3$.[4] With those restrictions we can identify an unique equilibrium where $T \in [2;9]$ adopt the modified strategy and $T \in \{1, 10\}$ rely on the original fundamental strategy. The return distribution in equilibrium is plotted in Fig. 5 and Table 2 shows the returns in equilibrium compared to the returns derived when an agent leaves equilibrium and adopts the alternative strategy. Note, that losses for the low-informed agents decrease markedly by applying the modified fundamental strategy. Returns for the insiders also decline, showing that they get fewer opportunities to exploit the other agents.

Table 2 The first line shows the return distribution in equilibrium when $T = 1$ and $T = 10$ apply the original fundamental trading strategy and all other informed agents apply the modified fundamental trading strategy as specified by equation 5. The second line shows returns emerging when one agent changes his strategy while all other agents stay with the equilibrium strategies

T	1	2	3	4	5	6	7	8	9	10
R_T equilibrium	−0.13	−0.18	−0.18	−0.14	−0.05	0.04	0.09	0.11	0.11	0.33
R_T alt. strategy	−0.16	−0.20	−0.26	−0.27	−0.27	−0.23	−0.16	−0.05	0.07	0.17

[4] First results from Table 1 as well as preliminary optimizations indicate that a value of 3 is the best common choice for most agents if we allow only integer values for Z.

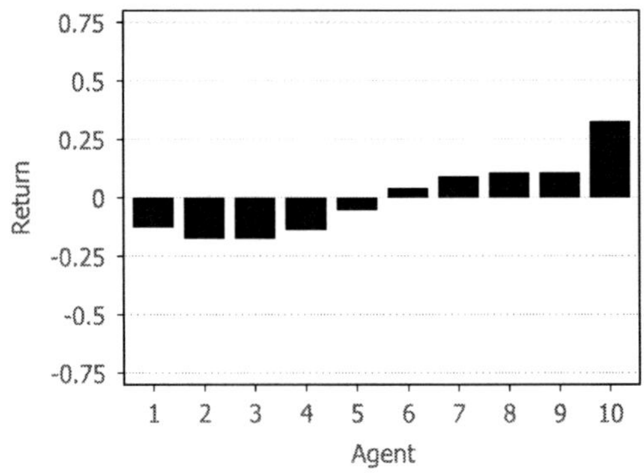

Fig. 5 Average returns for all ten agents in equilibrium when agents $T = 1$ to $T = 8$ adopt a modified fundamental strategy as specified by equation 5

When agents $T \in [2;9]$ apply the modified fundamental strategy, the correlations of $D_{T;r}$ with the signals change markedly (see Fig. 6). As expected, the positive correlations with the marginal signals are now much higher for all agents. The buy and sell orders of agents T_8 and T_9 are perfectly correlated with the agents' marginal signals. In contrast, the negative correlations with the signals unknown to an agent are less pronounced, indicating that herding is partly avoided.

This result is supported by the finding that market inefficiency decreases to a value of $\sigma_M^2 = 0.18$ in equilibrium. Market prices now reflect the fundamental value of the contract better than in the situation where all agents adopt the original fundamental strategy, as well as in the situation described in Sect. 4.2 where only one agent has been optimized.

5 Conclusion

We extended a model proposed by [11] by introducing a modified fundamental trading strategy. By giving more weight to the marginal piece of information an agent receives, this strategy prevents agents from making joint mistakes with other market participants and therefore reduces the herding effect. Results show that less-informed agents in our model can profit from this trading strategy at the expense of the insiders. A practical implication of this finding is that traders conducting fundamental analysis might be better off when ignoring information that is also known to the vast majority of market participants. Naturally, the caveat of this implication is that traders on capital markets will hardly be able to identify the "marginal" or most exclusive information they have. Empirical research on this issue might derive interesting results.

As a second result we showed that informational efficiency of market prices improves when agents place emphasis on their marginal piece of information. The

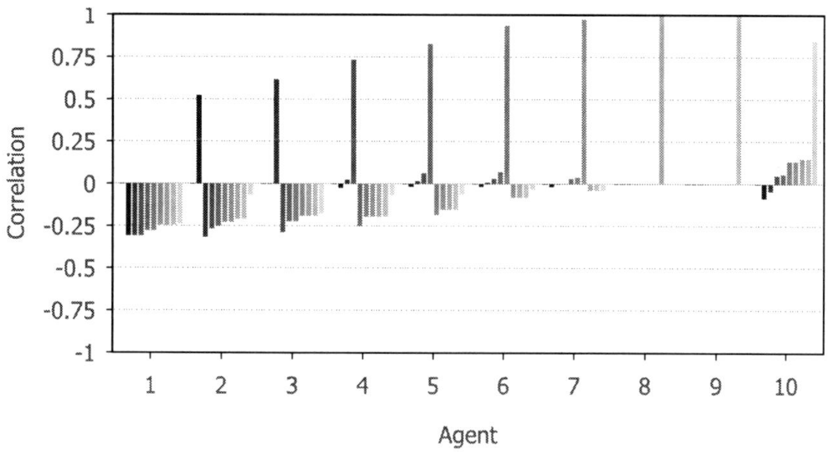

Fig. 6 Correlations of the $D_{T;r}$ variable of all agents with the individual signals ε_i when agents adopt the alternative fundamental trading strategy. Black bars denote correlations with ε_1 while the light grey bars denote correlations with ε_{10}

reason for this can again be attributed to the herding effect: When several agents trade on the same piece of information, market prices may overreact and thus become biased. When herding is avoided, market prices are more accurate with respect to the fair value of an asset.

As a last result we described an equilibrium when agents are bound to choose between a fundamental trading strategy and one version of the modified fundamental strategy that avoids herding. We showed that in equilibrium, the modified fundamental strategy is a dominant choice for eight out of ten agents, and market prices reflect the fundamental value of the contract to a large extent. The discussion of equilibria emerging when all agents can optimize their trading strategies without any restrictions will be left for future research.

Acknowledgements We thank Michael Hanke, Jürgen Huber, Klaus Schredelseker, and three anonymous referees for very helpful comments on this paper. We also acknowledge the FNR (Fonds National de la Recherche Luxembourg) for financial support on this project.

References

1. DeBondt, W.F.M., Thaler, R.H. (1985) Does the stock market overreact? Journal of Finance **40**: 793–808
2. Fama, E.F. (1970) Efficient capital markets: A review of theory and empirical work. Journal of Finance **25**: 383–417
3. Figlewski, S. (1982) Information diversity and market behavior. Journal of Finance **37**: 87–102
4. Hanke, M., Schredelseker, K. (2009) Index funds should be expected to underperform the index. Applied Economics Letters, Online First, doi:10.1080/17446540802599689

5. Hauser, F. (2008) On rational noise trading and market impact. In: Consiglio, A. (ed.), Artificial Markets Modeling. Springer, 141–153
6. Huber, J. (2004) A contribution to solving an old puzzle: why different trading strategies persist in competitive markets. Journal of Academy of Business and Economics **3**: 171–190
7. Lawrenz, J. (2008) Understanding non-monotonic payoffs for heterogenously informed agents. In: Hanke, M. and Huber, J. (eds.), Information, Interaction and (In)Efficiency in Financial Markets. Linde, 115–134
8. Lux, T. (1995) Herd behavior, bubbles and crashes. The Economic Journal **105**: 881–896
9. Pfeifer, C., Schredelseker, K., Seeber, G.U.H. (2009) On the negative value of information in informationally inefficient markets: calculations for large number of traders. European Journal of Operational Research **195**: 117–126
10. Schredelseker K. (1997) Zur Ökonomischen Theorie der Publizität. In: Ott, C., Schäfer, H.-B. (eds.), Effiziente Verhaltenssteuerung und Kooperation im Zivilrecht, J.C.B. Mohr, 214–245
11. Schredelseker, K. (2001) Is the usefulness approach useful? Some reflections on the utility of public information. In: McLeay, S., Riccaboni, A., (eds.), Contemporary Issues in Accounting Regulation, Kluwer Academic Publishers, 135–153

Stylized Facts Study through a Multi-Agent Based Simulation of an Artificial Stock Market

Zahra Kodia, Lamjed Ben Said, and Khaled Ghedira

Abstract This paper explores the dynamics of stock market from a psychological perspective using a multi-agent simulation model. We study the stock market trading behavior and the interactions between traders. We propose a novel model which includes behavioral and cognitive attitudes of the trader at the micro level and explains their effects on his decision making at the macro level. The proposed simulator is composed of heterogeneous Trader agents with a behavioral cognitive model and the CentralMarket agent matching buying and selling orders. Simulation experiments are being performed to observe stylized facts of the financial times series and to show that the psychological attitudes have many consequences on the stock market dynamics. These experiments show that the modelization of the micro level led us to observe emergent socio-economic phenomena at the macro level.

1 Introduction

The complexity of financial market still represents a wide field of research and applications. Many approaches are used to describe and to understand the market dynamics. Empirical and numerical analysis of stock market are powerful; nevertheless they still insufficient. Previous researches are generally based on the hypothesis of rational behavior. We distinguish an evolution of approaches used to study the stock market: (1) numerical approach during the eighties based on a stochastic modeling as [4], (2) multi-agent based systems simulating stock market dynamics (eg. [1], [5] [10]) and (3) behavioral multi-agent based systems in recent years: [9]. We note that psychological biases prevent traders from acting fully rational way and thus undermine the basic premise of the efficient market hypothesis. To deal with the complexity of the stock market, we use the agent modeling approach which offers

Zahra Kodia · Lamjed Ben Said · Khaled Ghedira
LI3/ENSI, Tunisia
e-mail: {zahra.kodia,lamjed.bensaid,khaled.ghedira}@isg.rnu.tn

M. LiCalzi et al. (eds.), *Progress in Artificial Economics*, Lecture Notes in Economics and Mathematical Systems 645, DOI 10.1007/978-3-642-13947-5_3,

the possibility to represent and to study two granularity levels of the market: the micro and the macro level. At the micro level, it gives the opportunity to observe the individual behavior of each actor. This approach allows to describe and to give for each entities the basic mechanisms to decide for itself when, how and why a particular action must be made [6]. At the macro level, the agent approach represents a powerful tool to observe the emergent results of interactive and cooperative systems.

In this paper, we propose a novel model representing the stock market and essentially based on cognitive behavior of the investors. Our artificial stock market includes the CentralMarket agent, mainly responsible for executing transactions via an order book, and a several kinds of Investor agents depending to their profile. We take into account three pairs of behavioral attitudes: (1) Pessimism / Optimism, (2) Speculation / Caution and (3) Mimetism / Leadership. This paper is structured as follows. The second section describes the micro/macro level of the stock market. First, we present the trader transaction protocol used in our simulation. Second, we detail investor decision making which is based on cognitive and rational paradigm and biased by behavioral attitudes. Then, we describe the structure of the used social network which reproduces the stock market dynamics. In the third section, we present experiments undertaken using our simulator. We study some stylized facts observed in our artificial market and we explain the behavioral effects on decision making. Finally, in the fourth section, we conclude and present future work.

2 The Micro/Macro Level of the Stock Market

2.1 Trader Transaction Protocol

We introduce a novel model representing the stock market dynamics. Our artificial stock market includes the CentralMarket agent essentially responsible for executing transactions via an order book. It includes also several kinds of trader agents depending from their profile. We take into account three pairs of behavioral attitudes: (1) Pessimism / optimism, (2) Speculation /Caution and (3) Mimetism / Leadership. Our artificial stock market is composed of: (1) a set of agents corresponding to the considered two types of actors: ExpertTrader agents and NoviceTrader agents, (2) a CentralMarket agent responsible of conducting transactions and controlling the dynamics of the stock market.

We describe now the trader's behavior while taking the decision of selling or buying and the interactions among the stock market actors. Figure 1 shows the Trader transaction protocol via an Agent UML sequence diagram. It specifies the sequence of messages that are exchanged along with their corresponding event occurrences on the actors' lifelines of our artificial stock market. Our two types of traders (expert and novice) have several features in common, we propose a generalization: the Trader agent. This sequencing diagram represents the confrontation of supply and

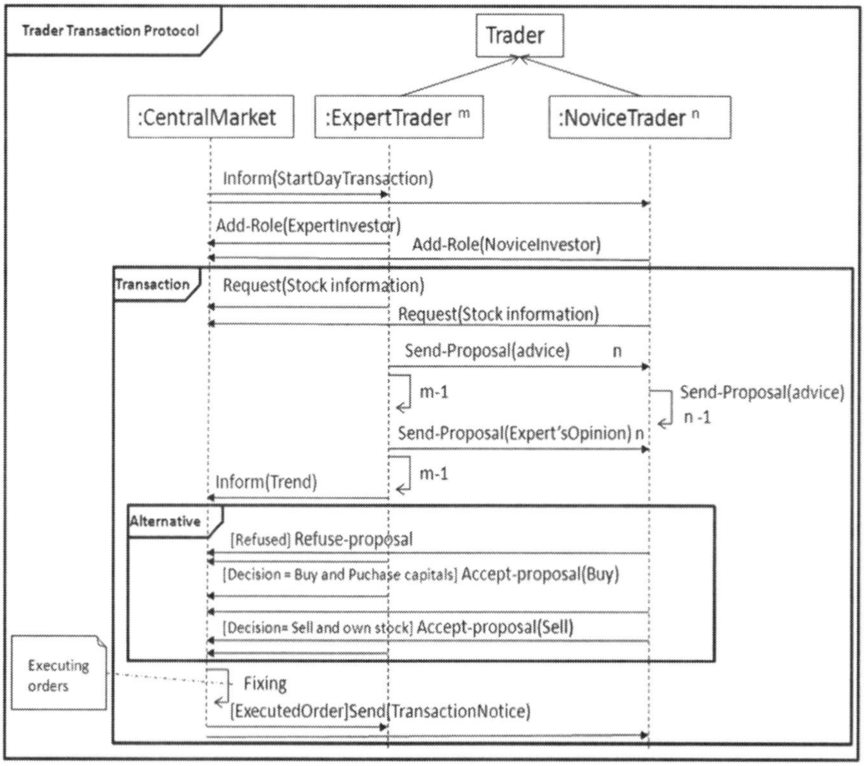

Fig. 1 Agent UML sequence diagram: Trader transaction protocol

demand for stocks by novice and expert traders. This confrontation takes place so that each actor transacts under his own criteria via the central market. The latter does not act as a centralized entity but as a meeting place used by traders to interact and realize stocks' exchanges. We use FIPA-ACL (Agent Communication Language) specifications to encode the messages' content. The first message is sent by the central market and informs traders that the transaction day is open. The add-Role message announces the presence of traders. It determines also the role played by the sender: ExpertInvestor role or NoviceInvestor role. ExpertTrader and NoviceTrader send a request message asking for stock information. CentralMarket replies for each message by collecting the needed information. If a trader gives a buy order, he must own in advance the capital corresponding to the amount of his purchase. In opposition, if he gives a sell order, he must own in advance the corresponding stocks. We assume that the transactions may include one or more stocks. After taking decision, ExpertTrader can either give advice or recommend a stock to other ExpertTraders and NoviceTraders through a Send-Proposal message. He communicates also the stock trend to the CentralMarket via an inform message. The NoviceTrader can also give advices to other trader playing NoviceInvestor role. Three choices are then possible for traders: (1) if the decision is to hold, trader refuses the proposal, (2) if the decision is to buy and the trader purchases stocks, he accepts buying and he sends a buy order via an Accept-Proposal message and (3) if the decision is to sell and the trader owns stocks, he accepts selling and he sends a sell order. We define a market order as a request to trade a specific quantity of a stock at any price. Every 30 seconds during a trading day, CentralMarket executes the fixing by matching sell and buy orders and calculating the market price of the considered stock. Then, a notification (send message) is transmitted to the correspondent senders of executed orders.

2.2 New Cognitive Investor's Model

The cognitive behavior model of the trader illustrated in Fig. 2 describes his perceptual, informational and decisional processes. It includes the behavioral attitudes which influence these processes.

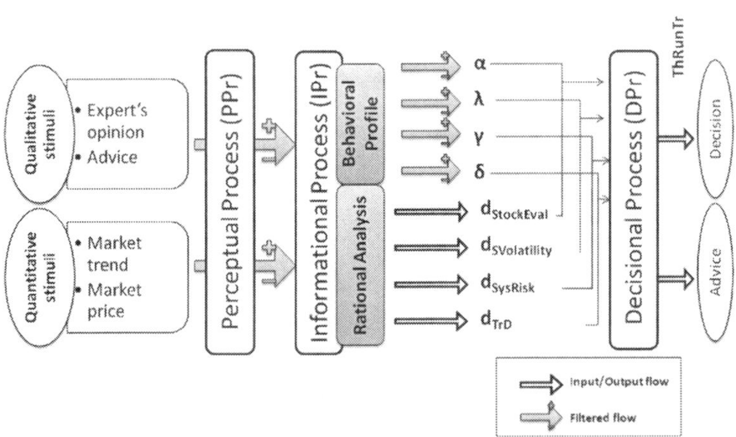

Fig. 2 The model of the investor cognitive behavior and his interactions within the stock market

Each Investor agent A_i is identified by an index $i \in (1,n)$ where n is the number of Investor agents. It is characterized by his $MarketExperience_i$ which represents its experience degree. We consider two investor experience levels : 0 for novice investor agent which has low experience and 1 for expert and professional investor.

The A_i behavioral profile, $Behavior_i$, is defined by a three dimensional vector ($OptimismV_i$, $SpeculationV_i$, $MimetismV_i$). Each component of this vector describes a behavioral attitude which can be either checked or reversed (represented respectively by +1 and -1). For example, A$_j$ who has its $Behavior_j \equiv$ (+1, +1, +1) is a optimist speculator mimetic investor, whereas if its $Behavior_j \equiv$ (-1, -1, -1), this agent is pessimist, cautious and leader. We note that all possible combinations are taken into consideration in our model.

Initially every A_i disposes of a number of randomly S_0 stocks and a fixed determined *Wealth*$_i$, the liquidity that it owns to invest in stock market. Each step of the simulation, based on quantitative and qualitative stimuli, the investor agent takes the decision of buying, selling or doing nothing and actualizes its *Portfolio*$_i$ (the set of stocks that it disposes).

2.2.1 Model Description

The cognitive and behavioral model that we propose is divided into three components. Figure 2 shows the links, the inputs and the outputs of these components.

- Perceptual process
  Trader agent takes into account the message received after checking it through the filter of privacy and / or the filter of confidence. For the privacy filter, if the sender of the message is part of trust network of the receiver, the message is accepted. Otherwise, the message is refused. For the confidence filter, if the acceptance threshold is lower than the information certainty threshold, this information is taken into account. Otherwise, the receiver refuses the message.
  Investors are influenced by their observation of the others choices without knowing whether it can be beneficial. In our model, advices can be considered although they are not certainly true. The purpose of the privacy filter consists in considering the sender of the message instead of its content. We use a dynamic network of trusted neighbors of the agent A_i, the CNeighborNet$_i$. This network is inspired from the six degrees of separation concept, [10], and is applied to the field of stock market neighborhood, [2]. We assume that the cardinality is equal to six Investor agents. On the privacy filter, if the sender of the message is part of CNeighborNet$_i$ of the receiver, the message is accepted. Otherwise, the message is refused.
  Confidence filter receives as input qualitative stimuli: expert's opinions and advices. Expert's opinion is a message transmitted by Investor agent having MarketExperiencei equals to 1 and defines the recommendation to buy or to sell a specific stock. Additionally, advice represents a stock judgment and it is transmitted by any Investor agent to its acquaintance. Interactions and relationship between Investor agents are defined by the Watts-Strogatz model of social network. Every qualitative stimulus while circulating in the market has a degree of conviction named CDegree. The qualitative information is taken into consideration whenever a confidence threshold is reached. The confidence threshold called ThConfidenceSensibility$_i$ is depending of the investor behavioral profile BP$_i$. If $CDegree < ThConfidenceSensibility_i$, then the relative message is taken into account. Else, the receiver refuses the message.
  We assume that the market trend and the market price are determinant and basic parameters in investor making decision. These two elements, which constitute

the quantitative stimuli, are neither affected by the investor profile nor by the perceptual process. Consequently, they are transmitted directly to the informational process. In addition, investor can give and receive advice or opinions from its neighbors. We take into consideration two qualitative stimulus: experts' opinion and advice.

The investor behavioral attitudes affect the perceptual process as shown in Fig. 2. In fact, the optimistic investor agent, which has a confidence in the outcome does not react the same way as the pessimistic investor agent. We assume that a speculator investor presents a lower $\text{Th\_ConfidenceSensibility}_i$ than a cautious investor who checks up a large number of received messages. Indeed, the number of persons composing CNeighborNet_i of an imitator investor is larger compared to the leader who has confidence in a few number of investors.

- Informational process

  Informational process treats qualitative and quantitative stimuli filtered by perceptual process. Investor agent ought to buy stocks which are deemed undervalued. It might sell the stocks which are considered overvalued. We consider in our model four tests related to fundamental analysis and chartist analysis: stock evaluation, stock volatility, systematic risk and trend determination. Each kind of tests provides a numerical signal d to buy, to sell or to hold (respectively +1, -1 and 0).

  The first analysis (stock evaluation) is based on the constant growth model known as Gordon-Shapiro model. This model only requires data from one period and an average growth rate which can be found from past financial statements [12].

  The second analysis is based on the risk measuring. We used the notion of systematic risk (or market risk), which is indicated by a given coefficient β. This coefficient indicates how the expected return of a stock or portfolio is correlated to the return of the financial market as a whole. It is calculated by the CentralMarket.

  The third analysis determines the growth rate of dividends. The latter is necessary in order to use the dividend discount model, which assumes that a stock's price is determined by the estimated future dividends, discounted by the excess of internal growth over the firm's estimated dividend growth rate.

  The fourth analysis is based on the hypothesis that the past development of a financial asset provides better information about its own future. In our model, we are guided by the Points and Figure Charting (PFC) [7]. This method represents the changes of the stock price and announces the signal of buying or selling.

- Decisional process

  The ExpertTrader agents' decision-making takes place after the four tests relative to: stock evaluation, risk measuring, dividend rate measuring and chart analysis. Each test gives out respectively a signal to buy, sell or do nothing respectively designed by $d_{StockEval}$, $d_{RiskMeasure}$, $d_{RDividend}$ and d_{TrD}. The final decision (as shown in Fig. 2) is calculated as follows:

$$D = \alpha * d_{StockEval} + \lambda * d_{RiskMeasure} + \gamma * d_{RDividend} + \delta * d_{TrD} \tag{1}$$

We notice that the parameters α, λ, γ and δ are generated randomly under the condition that their sum is equal to 1. Therefore, D ranges from -1 (which indicates buying) to +1 (which indicates selling) and just represents a signal. We note that D will be transformed to an order expect if $\|D\| \leq Th\_RunTr_i$. Th_RunTr_i defines a threshold for which the agent A_i run a transaction following its decision-making.

2.2.2 The Behavioral Attitudes

For the representation of behavior attitudes, we adopt the generic approach introduced in [3]. This approach is based on the specification of a set of inhibitor and triggering thresholds. In fact, each trader agent receives various kinds of qualitative stimuli (experts' opinion and advice) and quantitative stimuli (market trend and market price). The stimuli affect the trader agent decisions of buying or selling and play the role of reactive modulators that filter and weight the effect of external stimuli.

- Optimism / pessimism attitudes
  In our model, this behavioral component plays a crucial role in determining the estimated rates. The optimistic trader agent, which has a confidence in the outcome, does not react the same way as the pessimistic trader agent. These attitudes affect the rational analysis and more specifically the evaluating performance. Furthermore, in the informational process of our model, the optimistic trader agent over-estimates rates while the pessimistic trader agent under estimates them.
- Speculation/caution attitudes
  A speculator decides to conduct a transaction (buy or sell) by accepting the risk of losing in order to gain maximum benefits. On the contrary, a cautious trader agent proceeds with prudence and prefers to take every detail into account before buying or selling. This difference influences the rational analysis. Indeed, the speculator trader agent decides to buy a stock even if it presents a high rate of risk, while this would be unacceptable for the cautious trader agent. The parameters taken into account in PFC method related to the chart analysis are also influenced by this behavioral component. Besides, we consider that these attitudes affect the perceptual process. In fact, a speculator trader presents a lower threshold of acceptance information than a cautious trader.
- Imitation/leadership attitudes
  An imitator trader agent reproduces the reaction of other investor which represent his neighborhood. It follows and is aligned with the overall trend of the market. Whereas, the leader takes initiatives to buy or sell stocks. These behavioral attitudes influence the perceptual process which is more extended for the leader than the imitator trader. Indeed, the number of persons composing the confidence network of an imitator trader is larger compared to the leader one who has confidence in a few number of traders.

2.3 Social Networks and Interactions

In this section, we discuss the social interconnections in our artificial stock market. We underline the effects of the trader neighborhood on the decision making process and on the emergence of a global behavior. Indeed, behaviors of traders are related to the structures in which they fit. In our simulation, the neighborhood is not physical but it is a neighborly relationship (trust, privacy). A trader can make and receive advices or opinions from its neighbors. The links between traders are very complex and characterized by two effects, namely the clustering effects and the small world effects. High network clustering characteristic means that two individuals are much more likely to be friends with another if they have one or more other friends in common. The "small world" feature indicates that the average distance between individuals increases with the number of nodes in the networks logarithmically. We assume that the stock market is represented by a Watts-Strogatz (W-S) model [13] of social network. The W-S model fits very well on both small-word and clustering characteristics. We define social network as a social structure made of traders ("nodes") which are tied (connected) by one or more specific types of interdependency such as friendship, trust and privacy. This interconnection provides investors clusters which constitute the $CNeighborNet_i$ where information circulates randomly among heterogeneous set of traders. The boundaries of these domains change in time, corresponding to a situation where the links between individuals are dynamic only throughout the history of the network. At the beginning of the simulation, the $CNeighborNet_i$ is randomly chosen. Every step of simulation, each Investor agent A_i updates its $CNeighborNet_i$ by deleting the transmitter of a "wrong" advices (provides deficit). It replaces it by any randomly chosen Investor agent.

3 Experiments and Results

Our simulator is implemented using the MadKit platform [8] implemented with Java programming language. The fact that MadKit has specific environment modeling with an organizational architecture rather than specific agent architecture or specific model of interaction, responds well to our needs in modeling which is based on an organizational structure for investors. For all the experiments, we run the simulation during 25000 steps and with 200 Investor agents (divided into 80 expert investors and 120 novice investors). A W-S social network is defined where agents are interconnected with their 4 nearest neighbors and randomly with each possible other agent using 0.1 probability. This results in a random network where traders have 6 contacts on the average. Average length path is 5.8 and cluster coefficient equals to 35.27%.

We consider four stocks with random market prices. We notice that the upward movements are longer than downward one. This reflects a resistance to downtrend pressure during bull markets. Figure 3 shows the price fluctuations in our artificial market. In order to study autocorrelations in the measure of volatility, we represent the plot of the logarithmic returns ($ret(h) = \log p(h) - logp(h-1)$) illustrated in this figure. The model thereby introduces price volatility correlations which is an element of stylized facts.

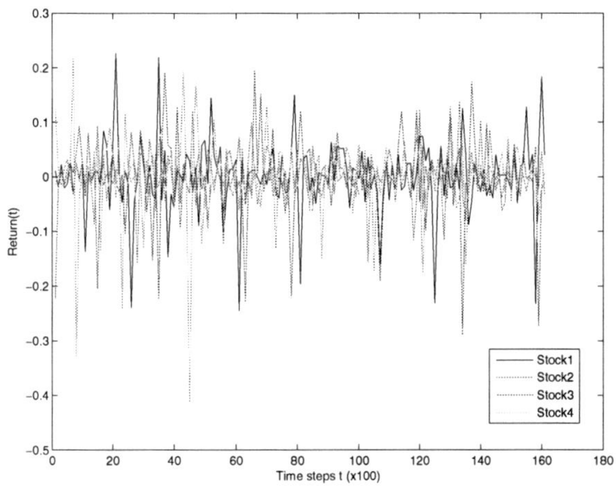

Fig. 3 The logarithmic returns of four stocks under our artificial stock market

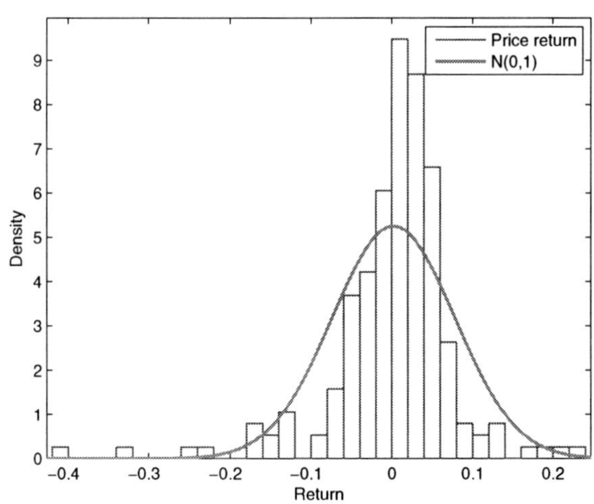

Fig. 4 Fat tails (Leptokurtosis) of the distribution of returns

We are typically concerned about whether the price return is distributed according to a normal distribution, since many of the statistical inference procedures that we use require the assumption of normality of the returns. Price returns exhibit fatter tails than the standard normal, or Gaussian, distribution and presents a kurtosis equals to 5.52. A remarkable Leptokurtosis behavior can be seen in Fig. 4. This figure presents the price returns distribution compared to a theoretical normal distribution having the same mean and variance.

Furthermore, financial time series usually exhibit a characteristic known as volatility clustering, in which large changes tend to follow large changes, and small changes tend to follow small changes. In fact, large shocks of either sign are allowed to persist, and can influence the volatility forecasts for several periods.

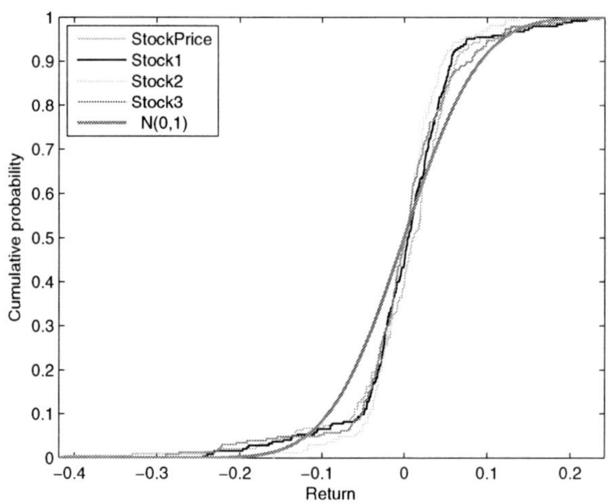

Fig. 5 Cumulative distribution function of Returns

Volatility clustering, or persistence, suggests a model where successive perturbations, although uncorrelated, are nonetheless serially dependent. Figure 5 plots the cumulative distribution function (CDF) of the four stocks return and shows a comparison between normal distribution and the CDF of returns of stocks. We observe a crossover to the normal distribution which happens for empirical financial data.

A normal probability plot represents useful tool for assessing of data. Many statistical approaches make the hypothesis that the underlying distribution of the data is normal, so this plot can provide some assurance that the hypothesis of normality is not being violated, or provide an early warning of a problem with your assumptions. We could notice that on Fig. 6 there is clear evidence that the underlying distribution is not normal.

Hence, our behavioral model is able to reproduce stylized facts observed in real stock market and to assure no predictability of future price developments and an efficient price formation.

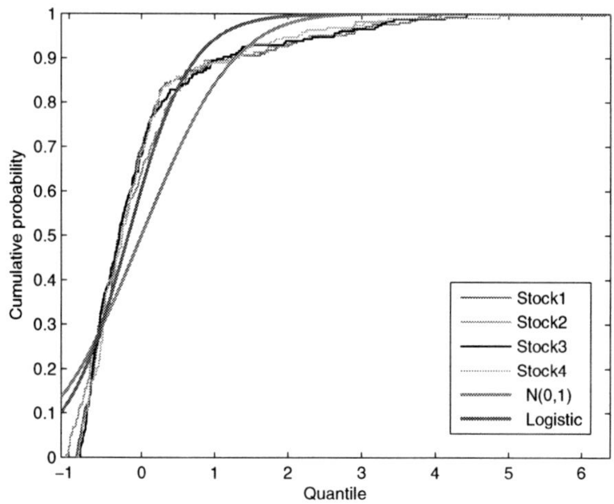

Fig. 6 Normal probability plot of the absolute value of return

4 Conclusion

Our research focuses on the modeling of the stock market trading through modeling the behavior and decision making of investor agent. Our contribution is to consider the stock market as a social organization of autonomous actors with dependents heterogeneous beliefs and different behavioral attitudes. Our model exhibits some stylized facts (i.e., volatility clustering and fat tails) and assures no predictability of future price developments and an efficient price formation. Different perspectives can be considered in our work. The first is to focus on the study of other stylized facts such as the Multi-scaling. The second is to explore the memory effect and the uncertainty at various levels in the stock market.

References

1. Arifovic J. (1996) The behavior of the exchange rate in the genetic algorithm and experimental economies. Journal of Political Economy **104**: 510–541
2. Bakker, L., Hareb, W., Khosravi, H., Ramadanovic, B. (2009) A social network model of investment behaviour in the stock market. Physica A: Statistical Mechanics and its Applications. **389**(6):1223–1229
3. BenSaid, L., Bouron, T. (2001) Multi-agent simulation of virtual consumer populations in a competitive market. In: Seventh Scandinavian Conference on Artificial Intelligence (SCAI01). Frontiers in Artificial Intelligence and Applications, IOS Press, Denmark, 31–43
4. Bray, M. (1982) Learning, estimation, and the stability of rational expectations. Journal of Economic Theory **26**(2):318–339

5. Derveeuw, J., Beaufils, B., Mathieu, P., Brandouy, O. (2007) Testing double auction as a component within a generic market model architecture. The Economy as a Complex Adaptive System, vol. 9 of Lecture Notes in Economic and Mathematical Systems. Springer
6. Dessalles, J., Muller, J.P., Phan, D. (2007) Emergence in multi-agent systems: conceptual and methodological issues. In: Phan, D., Amblard, F. (eds.) Agent-based modelling and simulation in the social and human sciences. The Bardwell Press, Oxford, 327–355
7. Dorsey, T. (2007) Point and figure charting: The essential application for forecasting and tracking market prices. Wiley Trading, 3rd Edition
8. Gutknecht, O., Ferber, J. (2001) The MadKit Agent Platform Architecture. In: Infrastructure for Agents, Multi-Agent Systems and Scalabile Multi-Agent Systems. LNCS 1887, Springer, 48–55
9. Hoffmann, A., Jager W., Von Eije J. H. (2007) Social Simulation of Stock Markets: Taking It to the Next Level. Journal of Artificial Societies and Social Simulation **10**(2):7
10. LeBaron, B. (2000) Agent-based computational finance: suggested readings and early research. Journal of Economic Dynamics and Control **24**:679–702
11. Milgram. S. (1967) The small world problem. Psychology Today **1**:60–67
12. Perold, A. (2004) The Capital Asset Pricing Model. Journal of Economic Perspectives **18**(3):3–24
13. Watts, D. J., Strogatz, S. H. (1998) Collective dynamics of ‘small world’ networks. Nature **393**:440–442

Part II
Auctions

A Variable Bid Increment Algorithm for Reverse English Auction

Imène Brigui-Chtioui and Suzanne Pinson

Abstract In this paper we propose multicriteria strategies for conducting automated reverse English auctions based on software agents. Reverse auctions gained popularity as a result of the emergence of Internet-based online auction tools. A buyer agent negotiates with several seller agents over a single product. The preference model is based on reference points which represent the desired values and the reservation values over each criterion. To insure process evolution, English auctions design often considers a bid increment that represents the minimal amount that a bidder must improve on the current best bid. Generally, the bid increment is fixed before the beginning of the process and kept invariant during the process. Our aim is to allow adjusting the bid increment as the auction process progresses. We propose an anytime algorithm based on an exponential smoothing method that adapts the bid increment to the auction context.

1 Introduction

From an economic point of view, auctions form an expanding study field from both the theoretical and the practical side [6]. Auctions have caught tremendous interest due to the well-defined negotiation environment they provide and the simple and clear rules they are based on. The many economic and computational aspects that need to be considered make the design of auction software a challenging task and have attracted considerable academic attention from the area of Artificial Intelligence, operations research and theoretical computer science [4]. An auction is

Imène Brigui-Chtioui
Institut Supérieur de Gestion, GRIISG, 147 Avenue Victor Hugo 75116 Paris
e-mail: imene.brigui-chtioui@isg.fr

Suzanne Pinson
Université Paris-Dauphine, Lamsade, Place du Maréchal de Lattre de Tassigny 75775 Paris
e-mail: suzanne.pinson@dauphine.fr

M. LiCalzi et al. (eds.), *Progress in Artificial Economics*, Lecture Notes in Economics and Mathematical Systems 645, DOI 10.1007/978-3-642-13947-5_4,

a negotiation process involving an auctioneer who conducts the process and sellers that are in competition in order to conclude the transaction.

In an English auction, the auctioneer begins by asking for bids at a low price and then gradually raises the price until only one willing buyer remains. The price increases along with the bidders bidding. When only one bidder remains in the auction process to bid for a certain type of sale item he or she becomes the auction winner. The theoretical literature [7] covers other auction environments too. We are interested in the English auction and more precisely reverse English auctions which imply one buyer or auctioneer negotiating with multiple bidders or sellers. Reverse auctions gained popularity as a result of the emergence of Internet-based online auction tools. They aim to assist the human buyer in the product choice.

In an English reverse auction, the buyer makes a counterproposal at each step which must be improved by the sellers at the next round. Otherwise, the sellers leave the auction. At each round, the best seller is kept waiting while the others receive a new counterproposal in order to improve the current best bid. The auction process goes on until all sellers (apart from the current best seller) leave the auctions or the time allowed to auction process is elapsed. English auction design often considers a bid increment that represents the minimal amount that a bidder must improve on the current best bid. Generally, the bid increment is fixed before the beginning of the process and kept invariant during the process. Our aim is to allow adjusting the bid increment as the auction process progresses. We propose an anytime algorithm that adapts the bid increment to the auction context.

The paper is structured as follows. After a related works section, we present our multicriteria auction model. We introduce in the fifth section the anytime counterproposal definition and we present its performance profile and its properties. We finally conclude and present our future work.

2 Related Works

Multiple emerging works focus on multicriteria negotiation [3, 10, 12] and especially on multicriteria auctions [1, 5]. Multiattribute auctions represent an extension to standard auction theory [9]. Their aim is to allow negotiating over all attributes of the product not only on price. Many approaches are proposed to allow negotiating over multiple criteria. The term of multicriteria auctions embraces many different approaches. A difference exists between approaches without an explicit scoring function and those using a scoring function. One of the multicriteria auctions of the first type is those of [13] which consists on a leap-frog method in which a bidder must make an improvement over the previous bid in at least one criterion and at the same time must behave not worse in any other criterion. The second type of multicriteria auctions is based on a scoring function that appreciates the overall utility of a bid. Thus, along the auction process, all attributes of the product can evolve. Utility is calculated by using an aggregation model that expresses the overall evaluation of a bid according to the value of the criteria that compose it.

The auction process starts when a buyer agent sends a request for proposal for an item to all interested seller agents. Sellers evaluate the request and send out acceptable bids or retire from the competition. Once the buyer has received all the replies, he selects the best bid and sends out a counterproposal based on the current best bid utility. The process goes on until at least one bidder is remaining. If it is the case, then the latter wins. In iterative auction mechanisms, the auctioneer provides some kinds of information feedback to the bidders to help them decide how they can improve their bids. In this article, we focus on the construction of the information feedback to the bidders which needs the formulation of the buyer counterproposal.

Whereas in price auctions only price could be improved by the seller (a win-lose situation), multicriteria auctions allow increasing the overall bid value considering all negotiable attributes. In this paper, we consider multicriteria English reverse auctions and we focus on the formulation of the buyer counterproposal. Most of the multicriteria auction models proposed in the literature [1, 2, 3] use a fixed increment operated on the current best utility to guarantee the auction evolution or leave the bidder free to fix the amount he improves on the current best utility. In the literature, we find two main counterproposal definitions in automated English auction mechanisms:

1. A fixed absolute increment is operated on the current best utility to set the next constraint [1, 6].
2. No increment is operated on the current best utility. In this case, the amount to improve on the current best utility is kept within the bidder province [3, 5].

In the first counterproposal definition, the bid increment is fixed along the auction process and doesn't depend on the current best bid utility. We consider that this bid increment definition have two main drawbacks. Firstly, fixing an absolute increment that represents the minimal amount to improve over a utility is significantly less intuitive than fixing a relative increment that represents the percentage to improve over this utility. For example, it's more natural to impose that bids should improve the best utility by at least 2% than 0.02 over the current best utility. Secondly, in a negotiation process and particularly in an auction process, negotiator's behavior should slightly change as the process goes along. The second counterproposal definition is not based on an increment but leave sellers free to decide which amount of utility they will improve over the current best bid. Thus, if all sellers make little increments over the current best utility, the process could last for a long time and consequently does not near the best bid. Moreover, these two counterproposals do not benefit from the auction context. We mean by auction context two main information: the remaining time and the number of bidders that are still in negotiation. This paper aims at addressing these drawbacks by proposing a counterproposal definition based on a variable relative increment. This increment is defined by an exponential smoothing method that uses the remaining time as key information to determine the relative increment at a given iteration. After presenting our algorithm, we demonstrate that it has the properties defined in [15].

3 Auction Protocol

In this section we present the multicriteria English reverse auction protocol. The protocol specifies the valid actions of each agent in a given context. In other words, the auction protocol specifies how the negotiation should be conducted. It defines rules about the sequence and the contexts of messages exchanged during an auction.

The system is based on a simple English auction protocol. The protocol specifies the valid actions of each agent in a given context. In other words, the auction protocol specifies how the negotiation could be conducted. It defines rules about the sequence and the contexts of messages exchanged during an auction [14]. The state graph in Fig. 1 illustrates the auction process which involves exclusively the buyer and the seller agents. At the beginning of the auction, the buyer agent sends a *CallForPropose* that indicates his preferences over the desired product. At each round t, the remaining seller agents propose bids that the buyer agent evaluates; this latter selects the best one, puts the best seller agent waiting, defines the next constraint and sends it to the remaining sellers. Either the seller agent proposes a bid that respects the current constraint imposed by the buyer agent, or he aborts.

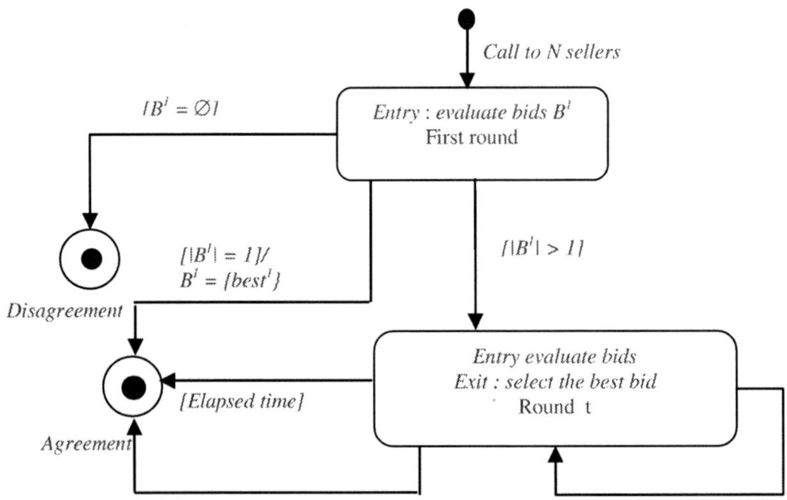

Fig. 1 Reverse english auctions

4 Auction Model

We start by proposing some preliminary useful notations:

- Let p, the number of attributes.
- Let $D = D_1 \times \ldots \times D_p$, the decision space where D_j is the domain of values for attribute j $(j = 1, \ldots, p)$.
- Let $C = C_1 \times \ldots \times C_p$ defines the criteria space.
- Let v_j, the value function defined from D_j to $C_j = [0;100]$ that corresponds to attribute j.
- Let $x = (x_1, \ldots, x_p) \in D$ denotes a bid, $b = (b_1, \ldots, b_p) \in C$ where $b_j = v_j(x_j)$ denotes the bid evaluated on all criteria.
- Let $\hat{t}$, the auction deadline.

4.1 Multi-Agent Reverse English Auction Architecture

The system presented in Fig. 2 consists of a buyer agent and a set of seller agents. Agents are connected via a web-based interface with end users. Every agent has a communication module that enables message exchange and treatment. For the buyer agent, the decision module is in charge of the counterproposal definition based on the seller agents' propositions and the strategy that it selects. This module is in charge, for the seller agent, of constructing the proposition according to the buyer request and to its propositions via the inference module that selects the set of valid propositions.

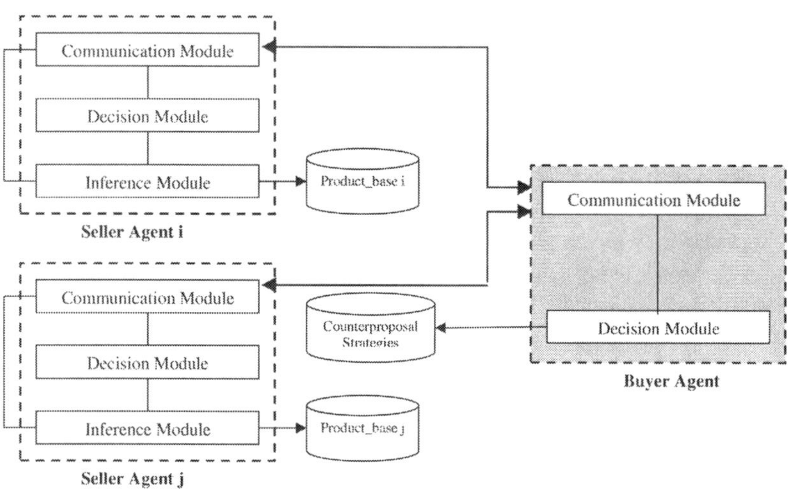

Fig. 2 Multi-agent reverse english auction architecture

4.2 Preference Model

The preference model is based on two reference points:

- The aspiration point, denoted by $a(a_1,\ldots,a_p)$ whose coordinates $a_j = v_j(dv_j)$ are aspiration levels, where $dv_j \in D_j$ is the desired value of the buyer on criterion j. Note that the aspiration point remains the same and is kept private during the auction process. The buyer agent formulates its counterproposal basing on its aspiration levels on each criterion but never communicates it.
- The reservation point, denoted by $r(r_1,\ldots,r_p)$ whose coordinates $r_j = v_j(mv_j)$ are reservation levels, where is the minimal value required on criterion j. The reservation point evolves during the auction process and is public to all seller agents. It enables seller agents to improve their propositions according to the buyer preferences.

4.3 Aggregation Model

The aggregation model determines the utility associated with a bid according to the buyer aspiration point. It is defined by the deviation from the aspiration point. This deviation measures the maximum difference between the aspiration levels and the bids values on each criterion. Equation 1 gives the utility expression of a bid b according to the aspiration point a.

$$U_a(b) = \max_{j=1,\ldots,p} \{(a_j - b_j)\} \tag{1}$$

Note that the multicriteria aggregation model is not a compensatory one. The model calculates differences between the aspiration point value and the bid value on each criterion and retains the higher difference as overall utility. Thus, a bad value on a criterion cannot be compensated by good values on other criteria.

We choose this multicriteria model because it addresses the drawbacks of the classical and well-known weighted sum. Firstly, it requires the definition of weights which are difficult to obtain and to interpret. Slight variations on these weights may change completely the choice of the best bid. Secondly, it can be shown that some of the non-dominated solutions cannot be obtained as the best proposal using the weighted sum for all possible choice of weights. This is a severe drawback since these non-dominated bids, whose potential interest is the same as the other non-dominated solutions, are rejected only for technical reasons.

While the buyer aims to reduce the deviation between bids and aspiration point, the preference relation based on the utility is given by Eq. 2.

$$b \succ b' \Leftrightarrow U_a(b) < U_a(b') \tag{2}$$

According to this preference relation, the buyer agent tries to minimize the deviation from the aspiration point. If there are more than one bid having the same best utility (called set $B*$), the best bid is selected according to a lexicographic algorithm: for each bid of $B*$, compute the deviation from the aspiration point a without considering the criterion on which the deviation is reached and determine the new set $B*$ of bids whose criterion values minimize the deviation. Repeat the process until $B*$ contains only one bid which is recognized as the best one, or until all criteria are eliminated, in which case the best is selected randomly. The use of this lexicographic algorithm ensures that the best bid is non-dominated. More details can be found in [1].

5 Anytime Counterproposal Definition

In the classical auction model [1], the constraint at t is given by Eq. 3. This increment is an absolute one and it remains the same along the auction process.

$$\forall t, U_a(b^{t+1}) \leq U_a(b^t) - \varepsilon \tag{3}$$

The counterproposal definition is presented in Fig. 3. $I_1 \bigcup I_2$ represents the set of propositions with same utility that the best proposition at round 1 (b_2). $I'_1 \bigcup I'_2$ represents the set of propositions with utility equal to $U_a(b_2) - \varepsilon$ that corresponds to reservation point e_2. At round 2, sellers' propositions will be in the zone delimited by $I'_1 \bigcup I'_2$.

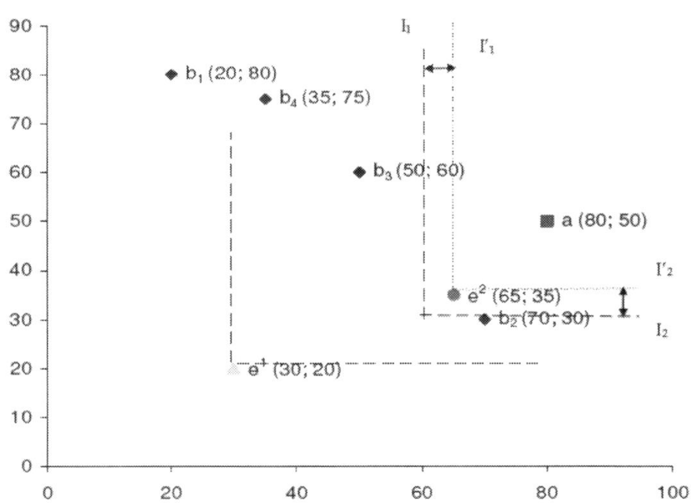

Fig. 3 Counterproposal definition

We propose an anytime counterproposal definition algorithm that insures an increment evolution and that satisfies the "beat-the-quote" rule introduced in [14] that leads to the following constraint:

$$\forall t, U_a(b^{t+1}) \leq U_a(best^t) \tag{4}$$

5.1 Propositions

In order to insure the evolution of the auction, a bid increment is defined over the utility of the current best bid. Our proposed technique is inspired from the exponential smoothing method that is a very popular economic method used to forecast the level of demand according to a current observation and previous observations. To employ this method, we consider two main observations at a given time t: the number of agent sellers that have not yet aborted $\mid B^t \mid$, and the remaining time $(\hat{t} - t)$. According to these observations, we formulate two propositions for the increment adjustment.

PROPOSITION 1. Using the exponential smoothing over the number of remaining sellers, the operated increment is calculated as follows:

$$\varepsilon^t = (\alpha \times \varepsilon^{t-1}) + ((1-\alpha) \times |B^t|) \tag{5}$$

where $\varepsilon^0 = 0$, $0 < \alpha < 0$, ε^t denotes the increment at round t and α denotes the smoothing factor.

The auction constraint at t is given by the Eq. 6:

$$\forall t, U_a(b^{t+1}) \leq U_a(best^t) - ((\alpha \times \varepsilon^{t-1}) + ((1-\alpha) \times |B^t|)) \tag{6}$$

PROPOSITION 2. We propose a new adjusting bid increment technique based on the remaining time as follows:

$$\varepsilon^t = (\alpha \times \varepsilon^{t-1}) + ((1-\alpha) \times (\hat{t} - t)) \tag{7}$$

where $\varepsilon^0 = 0$, $0 < \alpha < 0$, ε^t denotes the increment at round t and α denotes the smoothing factor.

According to the Eq. 7, the auction constraint at t is given by (8):

$$\forall t, U_a(b^{t+1}) \leq U_a(best^t) - ((\alpha \times \varepsilon^{t-1}) + ((1-\alpha) \times (\hat{t} - t))) \tag{8}$$

5.2 Properties

To allow anytime algorithms control, their performance improvement over time must be summarized quantitatively. The expected performance ξ is defined by evaluating the final agreement. We suppose as in [9] that disagreement is the worst outcome. Recall that disagreement happens when the condition $\mid B^1 \mid = 0$ is fulfilled. From this point of view, we establish:

$$\forall t, \xi(best^t) > \xi(Disagreement) \tag{9}$$

In the following, we present and demonstrate three properties that the proposed anytime algorithm satisfies [15].

PROPERTY 1. Measurable performance. According to this property, the quality of an approximate result can be determined precisely. We propose to measure the performance based on the utility of the winning bid. In order to maximize performance, we minimize the utility of the winning bid.

Proof. Considering two agreements and the corresponding winner propositions, we assert that:

$$\xi(g^1) > \xi(g^2) \iff U_a(best^1) < U_a(best^2) \tag{10}$$

Therefore, the proposed anytime mechanism satisfies the measurable performance property.

PROPERTY 2. Monotonicity. According to this property, the quality of the result is a nondecreasing function of time and input quality. The proposed anytime algorithm satisfies monotonicity.

Proof. A sufficient condition to insure monotonicity is to demonstrate that $(\alpha \times \varepsilon^{t-1}) + ((1-\alpha) \times (\hat{t} - t)) \geq 0$, which is equivalent to $\alpha \times \varepsilon^{t-1} \geq 0$ and $(1-\alpha) \times (\hat{t} - t) \geq 0$ that we can easily deduce from $0 < \alpha < 1$ and $(\hat{t} - t) \geq 0$.

CLAIM 1. Consistency. According to this property, the quality of the result is correlated to computation time and input quality. This property ensures to make a quality prediction of the algorithm.

To illustrate the third property, we represent the performance profile of Fig. 4 constructed experimentally on different instances of time. The performance profile represents the best utility reached at the end of the auction over time. In Fig. 4, we can easily see that, for the same input quality, the best utility decreases which means that the performance increases as the time allowed for the auction goes on. In this work, we mean by input quality the auction deadline, the best reachable utility among seller agents, the first imposed deviation and the number of seller agents.

To make a quality prediction of the proposed algorithm, we propose an algorithm that finds an approximation of the auction result according to a given input quality.

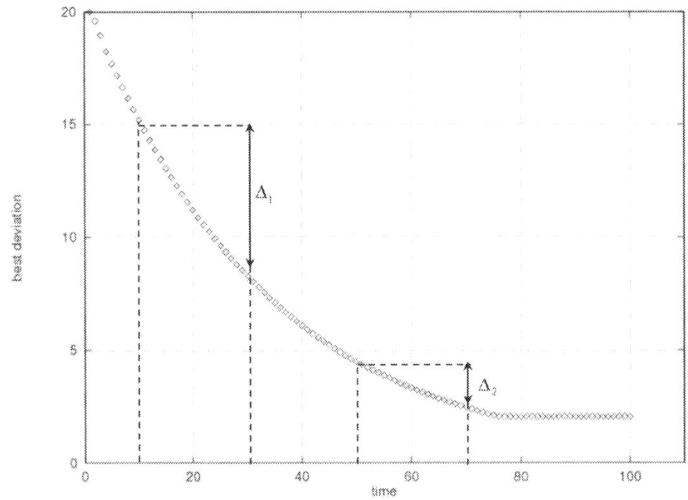

Fig. 4 Performance profile

The result achieved by the following algorithm is based on the assumption that all seller agents make bids that have a deviation equal to the deviation imposed by the buyer agent.

```
loop : while (t < t̂) and (d^t > best) do :
d^{t+1} = d^t - getIncrement(α, (t̂ − t));
setRemainingTime(); // updates the remaining time
end;
return lastConstraint();
```

CLAIM 2. Diminishing returns. According to this property, the improvement in solution quality is larger at the early stages of the computation, and it diminishes over time.

Figure 4 shows that at the beginning of the auction, the improvement in quality result is larger in the first iterations. For the same gap between the time allocated to auctions, the improvement performance is much greater in the first stages of the negotiation process (Δ_1) than in the ending stages (Δ_2).

6 Conclusion

In this paper, we proposed bidding strategies for the counterproposal definition in reverse English auctions. The counterproposal definition is based on an anytime algorithm that adjusts a variable relative increment. The increment is defined by an exponential smoothing method that uses the remaining time and the number of remaining sellers as key information to determine the increment at each step of the auction process. The proposed algorithm shows some interesting anytime properties like measurable performance, monotonicity, consistency and diminishing returns. In future works, we plan to undertake more validation, more specially to run the model with and without variable bid increment and to analyze the results. We also plan to experiment the algorithm properties in the case of other types of auctions like classical monoattribute auctions and multicriteria auctions based on compensatory aggregation models.

References

1. Bellosta, M.J., Brigui, I., Kornman, S., Vanderpooten, D. (2004) A multicriteria model for electronic auctions. 19th ACM Symposium on Applied Computing (SAC'04), Nicosia 759–765
2. Bichler, M. (2000) A roadmap to auction based negotiation protocols for electronic commerce. 33rd Hawaii International Conference on System Sciences (HICSS '00), IEEE Computer Society, Washington 759–765
3. Bichler, M. (2001) An experimental analysis of multiattribute auctions, Decision Support Systems **29**: 249–268
4. Bichler, M., Kalagnanam, J.R. (2006) Software frameworks for advanced procurement auction markets, Communications of the ACM **49**(2)
5. Bichler, M., Kaukal, M., Segev, A. (1999) Multi-attribute auctions for electronic procurement. In: First IBM IAC Workshop on Internet Based Negotiation Technologies, Springer 291–301
6. David, E., Azoulay-Schwartz, R, Kraus, S. (2003) Bidders' strategy for multi-attribute sequential english auction with a deadline. AAMAS
7. Klemperer, P. (2002) What really matters in auction design. Journal of Economic Perspectives **16**(1):169–189
8. Kraus, S., Wilkenfeld, J., Zlotkin, G. (1995) Multiagent negotiation under time constraints. Artificial Intelligence **75**:297–345
9. McAfee, R.P., McMillan, J. (1987) Auctions and Bidding. Artificial Intelligence **25**:699–738
10. Morris, J., Maes, P.: Sardine (2000) An agent facilitated airline ticket bidding system. 4th International Conference on Autonomous Agents (Agents 2000), Barcelone
11. Oliveira, E., Fonsesca, J.M., Steiger-Garao, A. (1999) Multi-criteria negotiation in multi-agent systems. 41st International Workshop of Central and Eastern Europe on Multi-agent Systems (CEEMAS'99), St. Petersbourg
12. Strecker, S. A. (2003) Preference revelation in multi-attribute reverse English auctions: a laboratory study. Twenty fourth International Conference in Information Systems
13. Teich, J., Wallenius, H., Wallenius, J. (1999) Multiple-issue auction and market algorithms for the world wide web. Decision Support Systems **26**:49–66
14. Wurman, P.R., Wellman, M.P., Walsh, W.E. (2002) Specifying rules for electronic auctions. AI Magazine **23**(3):15–23
15. Zilberstein, S. (1996) Using Anytime Algorithms in Intelligent Systems. AI Magazine **17**(3):73–83

Co-evolutionary Agents in Combinatorial Sealed-bid Auctions for Spectrum Licenses Markets

Asuncion Mochon, Yago Saez, Jose Luis Gomez-Barroso, and Pedro Isasi

Abstract Allocating scarce resources is a difficult duty governments must face. Furthermore, when participants exhibit complex preference structures (substitutes and complements) this task is even trickier. Combinatorial auctions are a good alternative for solving this problem. In this work we have developed a simulator of a combinatorial first-price sealed-bid auction. The bidding behaviour has been simulated by the application of co-evolutionary dynamics in an agent-based model. This model assumes independent bidders with bounded rationality trying to maximize profits. Finally, the simulations have been tested for two environments that involve the sale of spectrum licenses (*digital dividend*). These techniques are a helpful tool to support governments taking decisions in the awarding process.

1 Introduction

Combinatorial auctions are becoming a successful tool for allocating spectrum licenses and to determine the final price among competitors in many countries such as the USA, UK, Australia and Germany, among others. The market for spectrum licenses has the special characteristic that bidders' values for one license depends on the number of licenses already earned [2]. For one bidder, licenses can be substitutes (if earning one item reduces the value of earning more items) and complements (if already having one item increases the value of winning another one). Hence, the standard approach for allocating these items is by means of a combinatorial auction as it allows participants to bid both for complete packages of items as well as for individual items. Hence bidders can fully express their preferences [5].

Although combinatorial auctions are an excellent method for allocating resources with these properties, the awarding processes always involve a high level of uncertainty. There are many key decisions governments must face in auction design

A. Mochon · Y. Saez · J.L. Gomez-Barroso · P. Isasi
UNED and Universidad Carlos III de Madrid, e-mail: amochon@cee.uned.es

M. LiCalzi et al. (eds.), *Progress in Artificial Economics*, Lecture Notes in Economics and Mathematical Systems 645, DOI 10.1007/978-3-642-13947-5_5,

that can yield different outcomes. Hence, developing instruments able to give hints about potential results can be relevant. This paper presents a combinatorial auction simulator that can support governments taking decisions in this type of awarding processes. In this work we simulate the outcome of a first-price combinatorial sealed-bid auction and compare it to the efficient one. Furthermore, several scenarios have been simulated. To test the auction, we first developed specific software for solving this auction format. Then, to tackle the complex problem of simulating bidders' behaviour, a co-evolutionary system was developed, assuming bidders with bounded rationality. Evolutionary computation techniques are a useful tool for studying complex games like auctions, where the outcome depends on participants' valuations and strategies. These techniques gather evidence and support hypotheses for cases that have not been solved analytically. There is much research in which different computational techniques have been applied to auctions: [1], [3], [7] [18], [15], [22], [24], [27], and [30], among others.

The co-evolutionary computation design can be considered a special model of agent-based computational economics (ACE). It is an economic computational study modeled as dynamic systems of interacting agents [4]. In this paper, bidders are implemented as agents. Each bidder competes against the others for the lots that are being auctioned, and their aim is to maximize profits. The outcome of the auctions is assessed for two real-world scenarios related to the so-called digital dividend auctions in Europe. The first one simulates a case of limited (*national*) competition whereas a major foreign player participates in the auction in the second one. Results reveal that the auction yields more efficient outcome as competition increases.

The article is structured as follows. Section 2 describes the combinatorial first-price sealed-bid auction. Section 3 describes the scenarios tested: the radio spectrum to be auctioned and participants involved. In Section 4, we develop a competitive co-evolutionary system which allows us to simulate the complex behaviour of bidders in auctions. Section 5 deals with the results obtained and the last section is devoted to provide concluding remarks.

2 The Combinatorial First-Price Sealed-Bid Auction

In a combinatorial first-price single round sealed-bid auction, bidders are allowed to submit, in one single round, as many bids as they wish for any combination of available lots. For example, if we are auctioning licenses A and B, bidders might submit sealed bids for the following combinations: A, B and AB.

The formal expression of this auction format is the following. Let $M = 1,\dots,m$ be the set of items to be auctioned and $B_{j,i} = B_{j,1},..,B_{j,n}$ the set of package bids submitted by bidder j. A package bid is a tuple $B_{j,i} = (S_{j,i}, p_{j,i})$, where $S_{j,i} \subseteq M$ is a set of items and $p_{j,i} \geq 0$ the price bidder j is willing to pay for that package $S_{j,i}$. Then, the auctioneer determines the combination of feasible bids that maximizes his revenues, i.e., solves the Winner Determination Problem (WDP). A combination of bids is feasible if it allocates no item more than once. After that, the winners have to pay what they bid for the awarded items. Winning bidders can only win with a single bid, that is, bids are mutually exclusive (*XOR bidding language*). Solving the WDP is a NP-complete problem [26]. Hence, advanced computational techniques must be used to deal with this task. In this study, the A* based on a Branch On Bids (BOB) formulation search algorithm was used (see [25]).

To allocate radio spectrum licenses, the sealed-bid auction is often used as it is simple for bidders to understand and fast to implement. Moreover, in the presence of significant asymmetries among bidders and concerns about the level of competition, it promotes participation. The first combinatorial auction to assign a radio spectrum was held in Nigeria in 2002 [14]. The Nigerian Communications Commission (NCC) adopted a single round, first-price sealed bid combinatorial auction format. This same auction format has been used by other regulators in other countries, such as the Office of Communications (Ofcom) in UK.

3 The Scenarios

Nowadays, there are many countries involved in the process of digitizing television broadcasting signals. Transmission of existing television channels in a digital format requires a much smaller spectrum than does an analog broadcast. Thus, digital broadcasting makes it possible to free up a sizeable portion of the spectrum for new services in spite of new channels being added. These *newly* available resources are referred to as the *digital dividend.*

The radio electric spectrum is a scarce resource for which demand is growing quickly as the proliferation of innovative services that the spectrum requires to operate continues. Furthermore, the spectrum that will be released by digital switchover is in the sought-after UHF band. The UHF band offers a combination of coverage and capacity (bandwidth) that makes it suitable for virtually all common wireless applications. Therefore, allocation decisions for the digital dividend involve trade-offs between different potential uses. Most of the potential uses for the digital dividend are mass consumer services, in both television and mobile broadband. The European Commission is proposing the adoption of a set of common, coordinated actions, and in particular, the Commission is also considering a plan for harmonizing the 800 MHz band, remarkably suitable for new generations of mobile broadband.

Indeed, a growing number of major markets in Europe will use this band (comprising 72 MHz at 790-862 MHz) for mobile broadband. European countries are now setting out plans for their own digital dividends. To date, Finland, Sweden, France, Switzerland, Germany, Spain Denmark and the United Kingdom have decided to release the whole 800 MHz band, with others likely to follow.

This amount of spectrum has been (and will be) vacated and is being considered for release via beauty contests or, in many cases, via auctions. Our study simulates the auction of the 800 MHz band in a major-type European country. The total spectrum (72 MHz) is divided into 1 lot of 24 MHz, 2 lots of 16 MHz each and 2 lots of 8 MHz each.

There are many companies that would participate in such an auction of the spectrum allocated to mobile services. Mobile network operators licensed to provide digital (second generation) mobile services continue to dominate European mobile communications markets in spite of the presence of operators with UMTS (third generation) licenses, MVNO (mobile virtual network operators) and resellers.

The relative market shares of the leading operators, main competitors and newer entrants in each Member State have not changed significantly over the last few years. In most Member States, the two leading operators each have between 30% and 40% share of subscribers [9]. On average, in the European Union, the leading operator and its main competitor together account for 70% of the market share.

Considering this situation, two scenarios have been modeled:

- Scenario 1. Three agents participate in the auction. Two of them are considered strong (the leading competitor and its main competitor). The third one is a weak bidder: a company already present in the market but without the economic resources of the leaders, or a new entrant interested in offering services, deploying its own infrastructure.
- Scenario 2. Four agents participate in the auction. These are the same three as in Scenario 1 plus a mobile company operating in another country interested in entering this market. We consider this one to be a strong bidder as well.

The next step in simulating an auction is to determine what value bidders assign to the auctioned goods. The spectrum value for operators includes an engineering value and a strategic value: the engineering value reflects the less expensive configuration of the network infrastructure obtained when more spectrum is available to the operators network designers; the strategic value reflects the expected position in the market an operator will hold as a result of the assigned spectrum and the possible exclusion of another operator that would have otherwise come into play [28].

Both calculations are extremely complex: the first one requires engineering design exercises which are often quite particular and difficult to generalize. Even more complicated is the second one. In markets such as that of the ICT, the result will always be uncertain. However, the UK regulator's document detailing the proposed approach to the awarding of the digital dividend spectrum [19] contains assessments on the value that a certain additional fraction of spectrum could have for different services. With certain adaptations, they are valid for use in a simulation exercise such as the one performed here. In the model, the valuations are not a single figure but a range between a maximum and a minimum value which would correspond to different assessments (or confidence levels as regards the future profitability).

4 Bidding by Means of Agent-Based Co-Evolutionary Learning

Modeling bidders' behaviour in intricate games like auctions where bidders present both substitutes and complements among items, just as happens in spectrum license markets, is particularly complex. In these circumstances, where fitness values depend on the opponents' behaviour, the use of co-evolutionary learning techniques can became a valuable tool [11]. In fact, these techniques have been particularly successful for evolving robust competitive and cooperative behaviours in societies [16] and game learning problems [23] [31].

Co-evolutionary techniques arise from the idea of having a computer algorithm that learns from its own experience without being exposed to any human training. Apart from these techniques, the agent-based paradigm suits perfectly the modeling of the bidders' behaviour. Empirical and theoretical support for the application of co-evolutionary and agent-based models can be found in [29]. These techniques, considered a special model of ACE, have been useful for other similar problems such as oligopolistic market simulations [4].

The application of these learning techniques to auctions has an attractive property: the agents are implemented in a format that comes naturally; each agent is considered a developed species with an independent population of possible strategies, and the evaluation of each strategy (separate from the co-evolutionary system) depends on the other agents. This scheme allows us to have a pool of strategies that can act as a memory for agents between auctions ([31]; [17]) and that eliminates unhelpful cross-species mating.

The idea of introducing different species which evolve as natural ecosystems is not new ([11]; [12]; [20]; [10], among others). It is biologically inspired by the origin and the evolution of the species ([6]). However, many other interesting works about ecosystem interactions can be found in the literature (i.e., [13]). In this study, we propose using competitive co-evolutionary techniques to analyze two combinatorial auctions. As far as we know, this is the first time that these techniques have been applied to this field (combinatorial auctions). The results obtained yield some hints about possible outcomes using these types of auctions that may be useful to government agencies making their decisions about the awarding process of spectrum licenses. In the developed model we have assumed bidders with bounded rationality making their decisions based on the proposed competitive co-evolutionary learning system. A detailed description of the developed model for bidding behaviour is described below.

The co-evolutionary architecture implemented in this work is very similar to the one presented by [21], but in this case we always choose the best strategy as the representative agent. As every agent wants to maximize their profit, we have constructed a genetic algorithm (GA) which tries to maximize each agent's profit. To evaluate one agent, the quality of the population of strategies is measured against the opponents' best representative. There are many possible methods for deciding which representatives are most adequate for the evaluation process. In this work, we simply let the current best strategy from each agent (bidder) be the representative for the evaluations. Each agent is evolved by its own GA and communication between them is limited to the evaluation process with the strategy representatives of each agent. When evaluating the strategies within an agent, the representatives from the others remain fixed. The fitness of the different strategies is the agents' profit when playing a determined set of strategies (this is the phenotype). Since bidders are looking for the best possible strategies (profit maximizing), it is very improbable that cooperation between players will be found. Finally, the results obtained for analysis purposes are collected from the joint evaluation of the representatives of each agent, and the strategies produced by the co-evolutionary system let us test and analyze the proposed auction model.

5 Analysis of the Results

When allocating public resources, a primary goal to achieve is efficiency (the assignment that maximizes total value as lots are awarded to those that value them most). Furthermore, governments are interested in revenue maximization and also have equity goals. In this section, we compare the outcome obtained with the combinatorial first-price sealed-bid auction for two scenarios. The results are compared to the efficient outcome, this is, the outcome if all bidders bid their true value.

Table 1 Seller's income (standard error)

	Seller's avg. income (in M euros)	Δ respect to efficient outcome (%)
First scenario		
Sealed-bid outcome	749 (82.29)	−34.72%
Efficient outcome	1,147 (36.72)	
Second scenario		
Sealed-bid outcome	932 (58.25)	−20.75%
Efficient outcome	1,177 (27.71)	

As stated in Sect. 3, in the first scenario we have assumed three bidders (two strong and one weak) bidding for 72 MHz. In the second one, four bidders are involved in the radio spectrum auction (two strong, one weak and one not-established strong player). Table 1 shows the seller's income obtained with the sealed-bid auction for both scenarios. The results reveal that implementing the sealed-bid auction in the first scenario yields a reduction on the seller's income relative to the efficient outcome of 34.72%, while the reduction in the second scenario is 20.75%. Moreover, the standard deviation is also higher in the first scenario relative to the second one. The seller's income is also presented with Figs. 1 and 2. These figures corroborates previous finding, this is, in the second scenario: i) the seller's revenue increases and dispersion reduces; and ii) the final outcome is closer to the efficient one.

The intuition of these results is that in the second scenario, the new bidder increases competition, so the seller's income increases yielding an outcome closer to the efficient one. With this experimental research governments can have an idea of potential incomes to be obtained with this auction format. Governments can model different scenarios and compare outcomes. The results of this experiment reveal that, with this auction format, governments should encourage competition to increase revenues (the second scenario involves an increase of 24.43% in revenues with respect to the first scenario[1]).

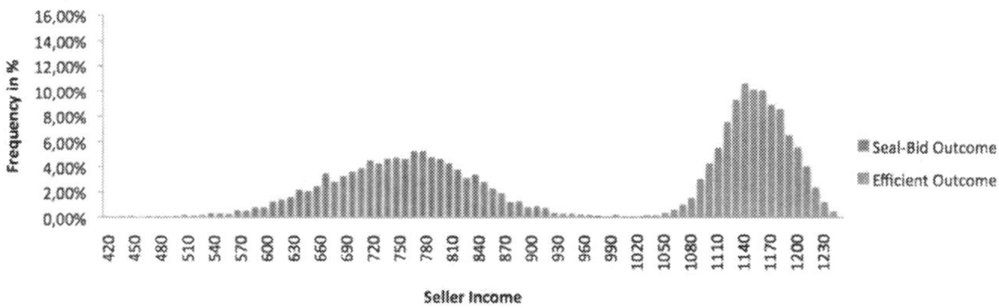

Fig. 1 Seller income histogram scenario 1 (Sealed-Bid vs Efficient Outcome)

Fig. 2 Seller income histogram scenario 2 (Sealed-Bid vs Efficient Outcome)

Another advantage of working with simulations in this environment is that governments can obtain hints not only about revenues but also about final allocation of the auctioned items. With this information governments can decide whether to change some variable from the auction design in order to achieve the main goals (efficiency, revenues maximization and equity). Furthermore, governments can determine new auctions rules in order to guarantee a specific service to improve social welfare. The spectrum allocation for the scenarios simulated in this work has also been analyzed, see Fig. 3 and Table 2.

The results reveal that as competition increases prices goes up and the weak bidder has fewer chances to earn items. In the first scenario bidder 3 (weak) earns 26.99% of the total spectrum auctioned, witch implies winning 24.81% over the efficient allocation. Nevertheless, in the second scenario this bidder only gets 4.42%.

[1] The difference between the seller's average income obtained for both scenarios is statistically significance for $\alpha = 0.05$, $t = -128.81$ and $p = 0.00$

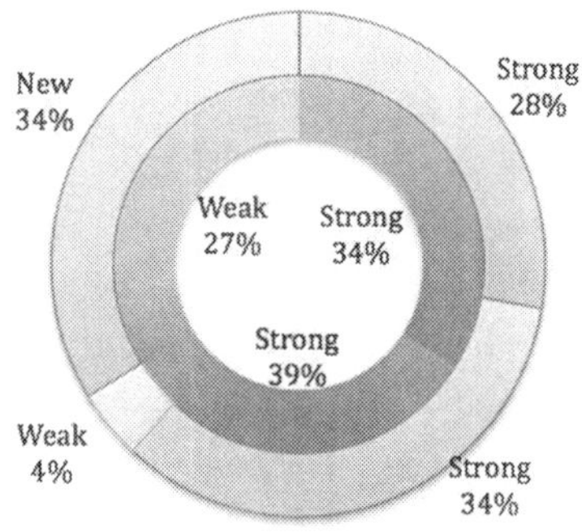

Fig. 3 Spectrum allocation for both scenarios. Inner ring: scenario 1. Outer ring: scenario 2.

The entrance of a new bidder yields a final distribution of the items closer to the efficient one among all bidders.

Table 2 Spectrum allocation in percentages of Mhz

	Bidder_1 (strong)	Bidder_2 (strong)	Bidder_3 (weak)	Bidder_4 (strong)
Scenario 1				
Sealed-bid	34.22%	38.79%	26.99%	N/D
Efficient allocation	43.92%	53.89%	2.18%	N/D
Scenario 2				
Sealed-bid	27.79%	34.00%	4.42%	33.79%
Efficient allocation	20.97%	49.33%	0.22%	29.47%

6 Conclusions and Future Work

Testing complex games like auctions is sometimes difficult to achieve with mathematical models. Nevertheless, the use of co-evolutionary computational techniques can help to deal with this task. This type of research gathers relevance when governments must allocate scarce resources, like the radio spectrum, in an efficient way, by means of auctions.

In this article, we have focused our attention on the combinatorial sealed-bid first- price auction. To test this auction format, a competitive co-evolutionary system for simulating the bidding behaviour has been developed. Each bidder is implemented as an agent searching for profit maximization competing against the others. Agents evolve according to a GA, and current best strategies are the representative for the evaluations. This model has been tested for allocating spectrum licenses assuming two scenarios with different intensities of competition among bidders. The first scenario involves three national bidders. In the second one there is a new bidder from another country. When comparing the outcome, the results match what was expected. The inclusion of a new foreign bidder makes the competition more intense resulting in an increase in the seller's income and a final outcome closer to the efficient one.

Translating these conclusions into policy recommendation, governments must foment bidders' participation as increasing competition moves auction results nearer to the efficient outcome. Nevertheless, both outcomes are still far from the efficient allocation. To dim this divergence governments could select another combinatorial auction format that reduces bidding shade and increases revenues. A good alternative could be the clock proxy auction, a combinatorial auction that starts with a clock phases and finishes with a proxy phase. Another solution could be to implement a different pricing rule. The Bidder-Pareto-Optimal core pricing rule developed by [8] is a good alternative as it is an approximate Vickrey-Clark-Grooves mechanism that yields core outcomes in any sealed-bid combinatorial auction. The main drawback of implementing any of these issues is that it increases complexity and makes it more difficult for bidders to understand the mechanism.

The main contribution of this work is the development of a simulator of a combinatorial auction as well as the implementation, by means of ACEs, of the bidding behaviour. Modeling or predicting potential outcomes for this type of awarding processes is not easy. Hence, governments are exposed to too many risks. Slight differences in the auction design or markets can yield important divergences in the final outcome. Therefore, the creation of this sort of tools means an improvement in future processes as governments can compare potential allocation and incomes based on simulations.

Acknowledgements This article has been supported by the Spanish research MCyT project MSTAR, Ref: TIN2008-06491-C04-03/TIN.

References

1. Andreoni, J., Miller, J.H. (1995) Auctions with Artificial Adaptive Agents. Games and Economic Behaviour **10**:39–64
2. Ausubel, L.M., Cramton, P., McAfee, P., McMillan, J. (1997) Synergies in Wireless Telephony: Evidence from the Broadband PCS Auctions. Journal of Economics and Management Strategy **6/3**:497–527
3. Avenali, A., Bassanini A. (2007) : Simulating combinatorial auctions with dominance requirement and loll bids through automated agents. Decision Support Systems **43**:211–228
4. Chen, H., Wang, X., Wong, K.P., Chung, C-Y. (2006) A Framework of Oligopolistic Market Simulation with Coevolutionary Computation. Lecture Notes in Computer Science **4221/2006**, Springer, Berlin/Heidelberg, 860–869
5. Cramton, P. (2007) How Best to Auction Oil Rights. In: Humphreys, M., Sachs, J.D., Stiglitz, J.E. (eds.), Escaping the Resource Curse, 114–151
6. Darwin, C. (1859) The Origin of Species, edition 1995, Gramercy

7. Dawid, H. (1999) On the convergence of genetic learning in a double auction market. Journal of Economic Dynamics and Control **23**:1545–1567
8. Day, R.W., Raghavan, S. (2007) Fair payments for efficient allocations in public sector combinatorial auctions. Management Science **53**(9):1389–1406
9. Commission Staff Working Document: Annex to the Communication COM (2009) 140. Progress Report on the Single European Electronic Communications Market 2008 (14th report).
10. Giordana, A., Neri, F. (1996) Search-intensive concept induction. Evolutionary Computation **3**(4):375-416
11. Hillis, W.D. (1990) Co-evolving parasites improve simulated evolution as an optimization procedure. Physica D. Nonlinear Phenomena **42**(1-3):228–234
12. Husbands, P., Mill, F. (1991) Simulated co-evolution as the mechanism for emergent planning and scheduling. In: Belew, R. K., Booker, L. B. (eds.), Proceedings of the Fourth International Conference on Genetic Algorithms, Morgan Kaufmann, San Mateo, 264–270
13. Lack, D.L. (1947) Darwin's Finches. Cambridge University Press, Cambridge
14. Kobodt, C., Maldoom, D., Marsden, R. (2003) The First Combinatorial Spectrum Auction. Lessons from the Nigerian Auction of Fixed Wireless Access Licenses, DotEcon DP 03/01:114–151
15. LiCalzi, M., Pellizzari, P. (2007) Simple market protocols for efficient risk sharing. Journal of Economic Dynamics and Control **31**(11):3568–3590
16. Lund, H.H. (1995) Specialization under Social Conditions in Shared Environments. Proceedings of the Third European Conference on Artificial Life ECAL'95, Springer-Verlag, 447–489
17. Mochon, A., Saez, Y., Quintana, D., Isasi, P. (2007) Bidding with memory in the presence of synergies: a genetic algorithm implementation. Proceedings of the IEEE Congress on Evolutionary Computation, 228–235
18. Mochon, A., Saez, Y., Isasi. P., Gomez-Barroso, J.L. (2009) Testing bidding strategies in the Clock-Proxy auction for selling radio spectrum: A Genetic Algorithm approach. Proceedings of IEEE Congress on Evolutionary Computation, 2348–2353
19. Ofcom (2007) Digital Dividend Review: a statement on our approach to awarding the digital dividend. United Kingdom Office of Communications. `http://www.ofcom.org.uk/consult/condocs/ddr/statement/`
20. Potter, M.A., De Jong, K.A. (1995) Evolving neural networks with collaborative species. In: Oren, T.I., Birta, L.G. (eds.), Proceedings of the 1995 Summer Computer Simulation Conference, The Society for Computer Simulation, San Diego, 340–345
21. Potter, M.A., De Jong, K.A. (2000) Cooperative Coevolution: An Architecture for Evolving Coadapted Subcomponents. Evolutionary Computation **8**(1):1–29
22. Reeves, D.M., Wellman, M.P., MacKie-Mason, J.K., Osepayshvili, A. (2005) Exploring bidding strategies for market-based scheduling. Decision Support Systems **39**:67–85
23. Rosin, C.D., Belew, R.K. (1995) Methods for competitive co-evolution: Finding opponents worth beating. In: Eshelman, L. (ed.), Proceedings of the Sixth International Conference on Genetic Algorithms, Morgan Kaufmann, San Francisco, 373–380
24. Saez,Y., Quintana, D., Mochon, A., Isasi, P. (2007) Effects of a rationing rule on the Ausubel auction: a Genetic Algorithm implementation. Computational Intelligence, **23**(2):221–235
25. Saez, Y., Mochon, A., Gomez-Barroso, J.L., Isasi, P. (2008) Testing BOI and BOB algorithms for solving the Winner Determination Problem in Radio Spectrum Auctions. In: Proceedings of the 8th International Conference on Hybrid Intelligent Systems, 732–737
26. Sandholm, T. (2002) Algorithm for Optimal Winner Determination in Combinatorial Auctions. Artificial Intelligence, **135**:1–54
27. Stone, P., Schapire, R.E., Littman, M.L., Csirik, J.A., McAllester, D. (2003) Decision-theoretic bidding based on learned density models in simultaneous, interacting auctions. Journal of Artificial Intelligence Research **19**:209–242
28. Sweet, R., Viehoff, I., Linardatos, D., Kalouptsidis, N. (2002) Marginal value based pricing of additional spectrum assigned to cellular telephony operators. Information Economics and Policy **14**(3):371–384

29. Tivnan, B.F. (2005) Coevolutionary dynamics and agent-based models in organization science. In: Proceedings of the 37th Conference on Winter Simulation (Orlando, Florida, December 04 - 07). Winter Simulation Conference. Winter Simulation Conference, 1013–1021
30. Wen, F.S., David, A.K. (2001) Strategic bidding for electricity supply in a day-ahead energy market. Electric Power Systems Research **59**(3):197–206
31. Yao, X., Darwen, P. (2000) Genetic Algorithms and Evolutionary Games. In: Barnett, W.A., Marks, R., Chiarella, C., Schnabl, H., Keen, S. (eds.), Commerce, Complexity and Evolution. Cambridge University Press, Cambridge, 313–333

The Effect of Transaction Costs on Artificial Continuous Double Auction Markets

Marta Posada and Cesáreo Hernández

Abstract Fast price convergence and high allocative market efficiency (close to 100%) are two of the most robust results in Experimental Economics. When human-subjects are replaced by artificial-agents, high allocative market efficiency is also attained even if the artificial agents have zero intelligence, but price convergence depends on the agents' learning. In this paper we study the sensitivity of Continuous Double Auction performance to the imposition of monetary costs in the market. We find that transactions costs reduce market efficiency. Price convergence results are very different when the monetary cost is imposed on the transaction or on the submissions to buy or to sell. Our agent-based market model confirms and extends previous Experimental Economics market results, and provides new behavioral explanations of the price dynamics.

1 Introduction

The Continuous Double Auction (CDA) is a double sided auction where buyers and sellers announce and accept offers to buy or to sell at any time. Although the valuations are private, fast price convergence and high allocative market efficiency (close to 100%) are two of the most robust results in Experimental Economics are two of the most robust results in Experimental Economics within CDA markets. Smith (1962) first demonstrated these properties, and subsequent researches have replicated them under alternative environment's conditions.

In a typical CDA experiment, transactions and submissions are costless. In this paper we study the sensitivity of the CDA market performance to the imposition of monetary costs using an agent-based modeling approach. Transaction costs are

---

M. Posada · C. Hernández
Escuela de Ingenierías Industriales, Universidad de Valladolid, Paseo del Cauce 59, 47011 Valladolid (Spain)
e-mail: posada@eis.uva.es, cesareo@eis.uva.es

M. LiCalzi et al. (eds.), *Progress in Artificial Economics*, Lecture Notes in Economics and Mathematical Systems 645, DOI 10.1007/978-3-642-13947-5_6,

relevant because in some markets costs of making offers exist as broker's commissions, travel costs of reaching markets, costs of writing contracts, and search costs of locating and identifying trading partners. The results are of theoretical and practical relevance because CDA is used in the real world trading of equities, CO2 emissions permits, derivatives, policy analysis like the Tobin's tax, etc.

Since the early Smith and Williams (1990)' work, little attention has been paid to the transaction costs effects on CDA markets. Our paper is related to Noussair *et al.* (1998)'s experimental work.

The major limitation of Experimental Economics is the lack of control for the human participant's behavior. When human-subjects are substituted by artificial-agents, the agents' behavior can be controlled. With artificial agents, the experimental results' robustness can be checked against alternative controlled agents' behavior with reliability and at low cost. Thus, Artificial Economics is the necessary companion of Experimental Economics, providing detailed explanations of the limitation and scope of Experimental Economics results.

In the last decade CDA markets have been extensively studied with artificial-agents. The results show that that high allocative market efficiency is achieved even if the artificial agents have zero intelligence (Gode and Sunder, 1993; LiCalzi and Pellizzari, 2008a, b). Both price convergence and individual surplus depend on the agents' learning (Posada *et al.*, 2006; Gjerstad, 2007; Posada and López-Paredes, 2008) explaining the paradox that a perfect market does not preclude intensive agents competition.

The purpose of this paper is to study CDA market performance when monetary costs are imposed on submissions to buy or to sell as in Noussair *et al.* (1998), and furthermore when they are imposed on transactions. We replicate and extend the results obtained in Experimental Economics and provide additional explanations of the transaction costs effects in the CDA market.

The structure of this paper is the following. In Sect. 2 we describe the microeconomic agent-based model. In Sect. 3 we calibrate the model and describe the experiments. In Sect. 4 we analyze the results in terms of market efficiency, price convergence and individual surplus when monetary costs are imposed to either submissions (to buy and to sell) or transactions and these results are compared to a CDA market with no monetary costs. Finally, in Sect. 5 we report the main conclusions of the paper.

2 The Agent-Based CDA Market Model

We consider a market where traders can exchange single units of a generic good. We describe our model in terms of the essential dimensions of any market experiment following Smith (1982): the institution (I) (the exchange rules, the way the contracts are closed, and the information network), the environment (E) (agent endowments and values, resources, knowledge) and the agents' behavior (A).

2.1 The Institution

The institution is a CDA. Any trader can submit or accept an offer to buy or to sell at any time during the trading period. There are several variations of the double auction exchange rules to simplify its implementation. LiCalzi and Pellizzari (2008a) pointed out that the simplifications of the CDA rules matter. We consider that in the market there are selling and buying books and the spread reduction rule is applied. Traders randomly place offers on the books. Orders are immediately executed at the outstanding price if they are marketable. Otherwise, they are recorded on the books and remain valid until either the end of the trading session (that is, without resampling) or the agent improves its offer (to buy or to sell).

2.2 The Environment

The environment is stationary (the competitive equilibrium price is the same in every period) in order to study, *ceteris paribus*, the price convergence performance.

Each trader is endowed with a finite number of units and a private valuation for each unit. Each agent has fifteen units to trade and their valuations are those reported in Noussair *et al.* (1998) in order to reproduce this experimental work (that is, 4 sellers and 4 buyers and a monetary cost on submissions to buy or to sell, TC1). Demand and supply are an approximation of $D(x) = 1535 - 30x$ and $S(x) = 35 + 30x$, respectively. Competitive equilibrium exits at any market price of between 780 and 790 and a quantity of 25 transacted. The consumer surplus is 9125 and the producer surplus is 9125.

In order to extrapolate the environment to a market populated by 12 sellers and 12 buyers, we consider an approximation of the following demand and supply $D(x) = 1535 - 10x$ and $S(x) = 35 + 10x$, respectively (which have been built by adding three equal demands and supplies, respectively). Competitive equilibrium exits at any market price of between 780 and 790 and a quantity of 75 transacted. The consumer surplus is 27375 and the producer surplus is 27375 (on the left of Fig. 1).

The monetary cost is of 50 units. We consider three alternative cases: when the monetary cost is imposed only on submissions to buy or to sell (TC1), when the monetary cost is imposed on transactions (TC2), and when there is no monetary cost (TC0). When a monetary cost is imposed, there is a loss of surplus, which reduces the allocative market efficiency.

This loss surplus can be calculated theoretically when the monetary cost is imposed on transactions (TC2). The competitive equilibrium exits at any market price of between 780 and 790 and a quantity of 69 transacted. The consumer surplus is 23805, the producer surplus is 23805, and the total monetary cost is 3450. Therefore, the ratio of market efficiencies is 0,932 (on the right of Fig. 1).

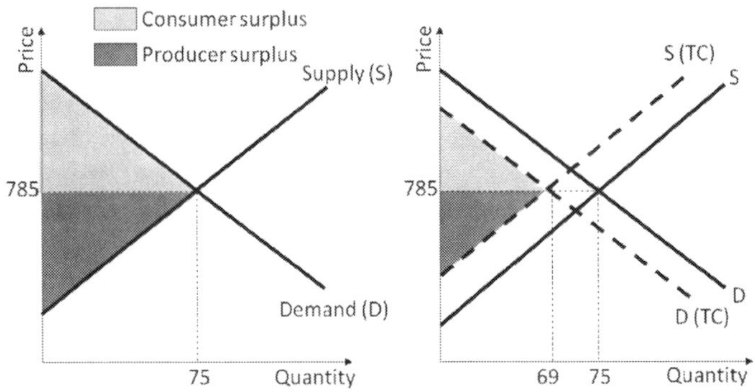

Fig. 1 Environment without monetary cost (left) and with monetary cost on transactions (right)

However, the surplus loss cannot be calculated theoretically when the monetary cost is imposed on submissions (TC1) that is why we turn to Experimental Economics research.

2.3 Agents' Behavior

Each trader is either a seller or a buyer. In CDA markets traders face three non-trivial decisions: *How much should they bid or ask? When should they place a bid or an ask? And when should they accept an outstanding order?* Bidding strategies corresponds to particular answers for these decisions. LiCalzi and Pellizzari (2008b) pointed out that learning greatly improves the expected value of the allocative efficiency in CDA without the resampling assumption. Therefore, we consider GD and K bidding strategies (whose parameters are detailed in Table 1).

GD is the most sophisticated learning designed for CDA (Gjerstad and Dickhaut, 1998) . The market performance is excellent in terms of price convergence and efficiency when the traders use a GD bidding strategy. Each agent chooses the offer that maximizes his expected surplus, defined as the product of the gain from trade

Table 1 Parameters of the bidding strategies and their default values

GD parameter	
Memory	8

K parameters	
Minimum profit	[0.01 , 0.03]
Time out	[0.05 , 0.15]
Ratio orders	[0.0125 , 0.0375]

and the probability for an offer to be accepted. GD agents modify this probability using the history.

Kaplan (K) is the simplest learning designed for CDA. It was the winner in the tournament of Santa Fe Institute in 1993 (Rust *et al.*, 1993) . The basic idea behind the Kaplan strategy is: "wait in the background and let others negotiate. When an order is interesting, accept it". K agents are parasitic on the intelligent agents to trade and to obtain profit. The market performance is poor in terms of price convergence and efficiency when most of the traders in the market use a K bidding strategy. However, its performance in terms of individual agents' profits is excellent. If all traders in the market are K agents no trade will take place.

3 The Experiments

We have analysed the following twenty-four scenarios (see Table 2) that accommodate a symmetric environment with:

- Different number of agents competing in the market (4 sellers and 4 buyers; 12 sellers and 12 buyers).
- Two kinds of learning agents (GD and K) with four alternative market populations: 100%GD-0%K, 75%GD-25%K (buyers), 62,5%GD-37,5%K (buyers), and 50%GD-50%K (buyers).
- Two sources of monetary costs: on submissions to buy or to sell, TC1, (per offer excise tax), and on each transaction, TC2, in order to compare the results with a no-monetary cost case, TC0.

E11, E21, E31, and E41 scenarios are the benchmarking scenarios, which allow us to compare Experimental and Artificial Economics results. They provide new explanations to those in the human based experiments.

Each run consists of a sequence of 10 consecutive trading periods, each one lasting 100 time steps.

Table 2 Simulated environments where monetary cots are on submissions (TC1), on transactions (TC2), and no-monetary cost (TC0)

Agents Learning	ENVIRONMENTS (sellers × buyers)					
	4 × 4			12 × 12		
	TC0	TC1	TC2	TC0	TC1	TC2
100%GD-0%K	E11	E12	E13	E14	E15	E16
75%GD-25%K (buyers)	E21	E22	E23	E24	E25	E26
62,55%GD-37,5%K (buyers)	E31	E32	E33	E34	E35	E36
50%GD-50%K (buyers)	E41	E42	E43	E44	E45	E46

4 Measures and Main Results

We analyzed CDA market performance in terms of price convergence and market efficiency. Our results confirm both expected theoretical effects and experimental results, and they provide new behavioral explanations of market dynamics.

4.1 Market Efficiency

As market efficiency is not a univocally concept defined in the economic literature, we make some comments about it. We deal with market efficiency as understood in conventional microeconomics, and following the experimental economics research on market dynamics. We define allocative market efficiency as the total profit actually earned by all the traders divided by the maximum total profit that could have been earned by all the traders (i.e., the sum of producer and consumer surplus) (Smith, 1962).

A comparison of the graphics of Fig. 2 shows that introducing monetary costs reduce allocative market efficiency in all the scenarios. We confirm expected theoretical results: the lowest efficiency in all scenarios is achieved when a monetary cost is imposed on each transaction (TC2). Our results also confirm experimental results: the efficiency is lower when a monetary cost is imposed on submissions (TC1) than when there is not monetary cost (TC0). Moreover, we obtain as well a measure of the relative effects of TC1 versus TC2, a relevant information when designing real CDA institutions.

However, the efficiency loss not only depends on the source of the transaction cost, but it also depends on the agents' learning. The efficiency decreases as the percentage of K agents increases in the market. There is a wide gap when this percentage goes from 37,5

In Table 3 the sample average and the sample standard deviation per run are reported.

Table 3 Simulated environments where monetary cots are on submissions (TC1), on transactions (TC2), and no-monetary cost (TC0)

Agents Learning	12 sellers × 12 buyers					
	Average			Standard Deviation		
	TC0	TC1	TC2	TC0	TC1	TC2
100%GD-0%K	0,9993	0,9337	0,8672	0,0002	0,0002	0,0002
75%GD-25%K (buyers)	0,9894	0,9153	0,8547	0,0037	0,0064	0,0050
62,55%GD-37,5%K (buyers)	0,9587	0,9027	0,8223	0,0131	0,0120	0,0151
50%GD-50%K (buyers)	0,3791	0,3784	0,3560	0,0072	0,0056	0,0045

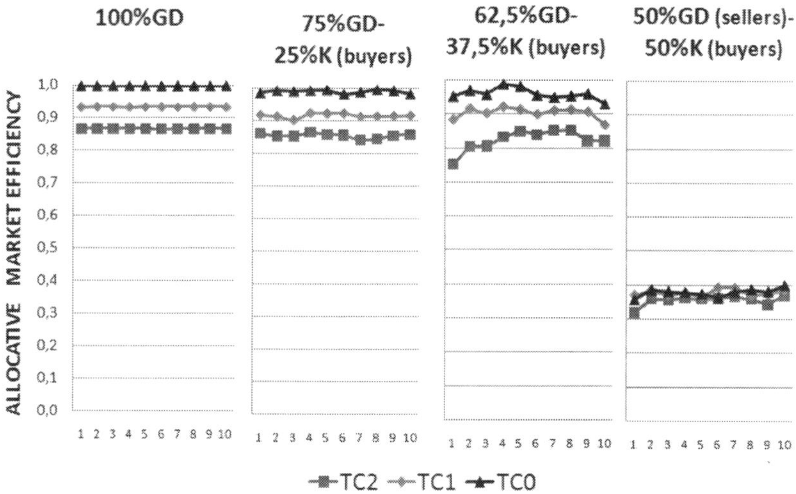

Fig. 2 Average efficiency per period of a market with 12 sellers x 12 buyers

4.2 Price Convergence

We find that price convergence is not always achieved. It depends on the agents' learning and on the source of the transaction cost.

Clearly, if one side of the market can produce such a price movement away from equilibrium, agents on that side of the market will benefit. Such price movements are frequent when the market is populated by K agents. (in other words, when there is parasitic agents in the market). This effect can be noticed by comparing transaction prices (compare Fig. 3 and Fig. 4).

Our results provide new behavioral explanations of the price dynamics. For instance, one of the unexpected results in Noussair *et al.* (1998) was that the number of asks submitted to the market by human subjects exceeded the number of bids. Their explanation was that, "because subjects have more experience as buyers than as sellers outside the laboratory, they may be more effective at reducing offer costs when acting as buyers than when acting as sellers in the laboratory". We provide another explanation due to the agents' behavior that can be controlled in Artificial Economics. In E21 to E46 scenarios, the number of offers submitted to the market by sellers exceeds the number of bids submitted by buyers. And, the extreme case is when only sellers submit offers to the markets as all buyers are K agents (that is in E41, E42, E43, E44, E45 and E46 scenarios). Our explanation rely on agents' learning skills.

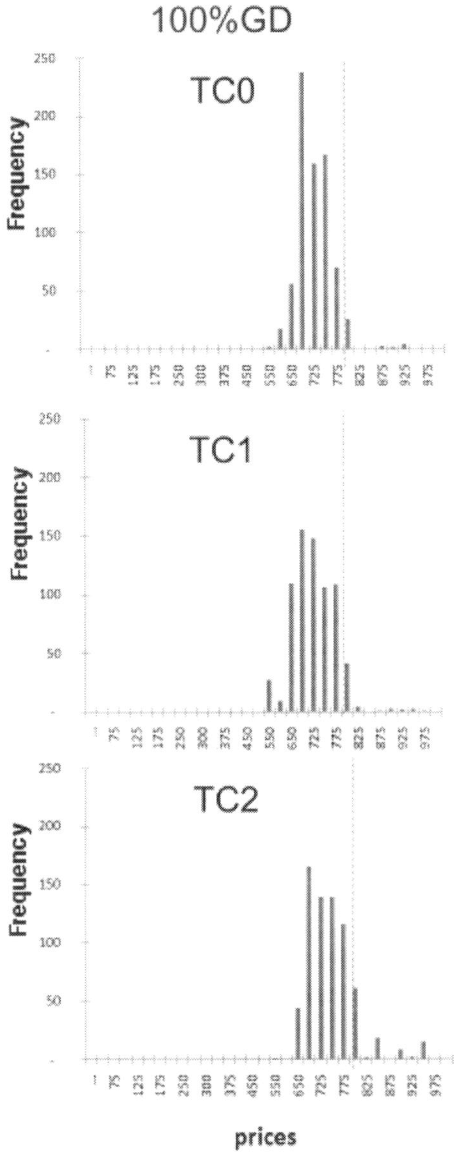

Fig. 3 Frequency distributions of price transactions when the market is populated by 100% GD agents and monetary cost: a) without, TC0 (E14), b) on submissions, TC1 (E15), c) on transactions, TC2 (E16)

5 Conclusions

Artificial Economics models are a necessary companion of Experimental Economic models. The CDA experiments with artificial agents have widened the scope and

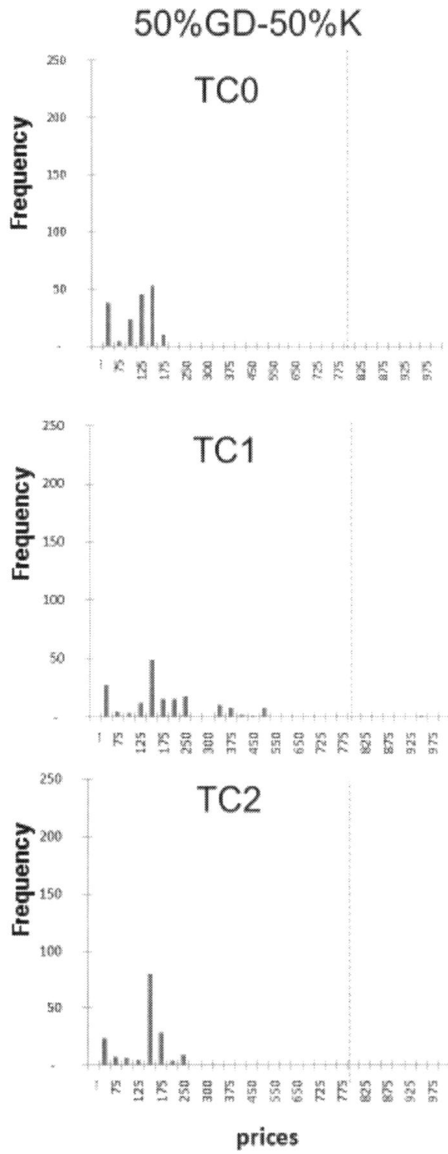

Fig. 4 Frequency distributions of price transactions when the market is populated by 50% GD and 50% K (buyers) agents and monetary cost: a) without, TC0 (E44), b) on submissions, TC1 (E45), c) on transactions, TC2 (E46)

they provide a thinner resolution and alternative explanations of price dynamics and social and individual efficiency for a wide range of microeconomic settings. When one adds realistic assumptions to the pure CDA, such as transaction costs and emotional agents, AE can be very useful to inspire explanations of the observed behavior. Our Artificial Economics model confirms and extends previous works on the CDA with transaction costs and provides new behavioral explanations of the price dynamics.

Posada *et al.*, (2006) first demonstrated that, if the market is populated by K agents, and more clearly if they are on one side of the market (making bids or asks), price convergence is not achieved. This result is even stronger when, as in our experiment, there are transaction costs which are imposed on the communications. The consequence for market policy design is that if monetary costs are imposed, it should be on the transactions, because communication costs reinforce the parasitic feature of the K agents.

We have assumed a symmetric environment to replicate the Noussair *et al.* (1998) experimental work, but a relevant extension could be (as in the CDA with no transaction costs) the study of the CDA price dynamics with asymmetric environments and endogenous changes in the bidding strategies.

Acknowledgements This work has received financial support from both *Junta de Castilla y León*: VA003B08, and *Ministerio de Ciencia e Innovación*: TIN2008-06464-C03-02.

References

1. Gjerstad S (2007) The competitive market paradox. J Econ Dyn Control 31:1753–1780
2. Gjerstad S, Dickhaut J (1998) Price formation in double auctions. Games Econ Behav 22:1–29
3. Gode D, Sunder S (1993) Allocative efficiency of market with zero-intelligent traders: Market as a partial substitute for individual rationality. J Polit Econ 101:119–137
4. LiCalzi M, Pellizzari P (2008a) Simple market protocols for efficient risk sharing. J Econ Dyn Control 31:3568–3590
5. LiCalzi M, Pellizzari P (2008b) Zero-intelligence trading without resampling. Lect Notes Econ Math Syst 614:3–14.
6. Noussair C, Robin S, Ruffieux B (1998) The effect of transaction costs on double auction markets J Econ Behav Organ, 36:221–233.
7. Posada M, Hernández C, López A (2006) Learning in a continuous double auction market Lect Notes Econ Math Syst 564:41–51.
8. Posada M, López-Paredes A (2008) How to choose the bidding strategy in continuous double auctions: Imitation versus take-the-best heuristics. JASSS 11(1)6 `http://jasss.soc.surrey.ac.uk/11/1/6.html`
9. Rust J, Miller J, Palmer R (1993) Behaviour of trading automata in computerized double auctions. In: Friedman and Rust (eds), The double auction markets: Institutions, theories and evidence. Addison-Wesley, Reading (MA)
10. Smith VL (1962) An experimental study of competitive market behavior. J Polit Econ 70:111–137.
11. Smith VL (1982) Microeconomic systems as an experimental science. Am Econ Rev 72:923–955
12. Smith VL, Williams A (1990) The boundaries of competitive price theory: Convergence, expectation and transaction costs. In: Green and Kagel (eds) Advances in behavioral economics 2:3–35. Ablex Publishing, New York

Part III
Networks

The Rise and Fall of Trust Networks

Kartik Anand, Prasanna Gai, and Matteo Marsili

Abstract The working of economies relies on trust, with credit markets being a notable example. The evaporation of trust may precipitate the economy from a good to a bad state, with long-lasting and large scale structural changes, witness the 2007/8 global financial crisis. Drawing on insights from the literature on coordination games and network growth, we develop a simple model to clarify how trust breaks down in financial systems. We show how the arrival of bad news about a financial agent can lead others to lose confidence in it and how this, in turn, can spread across the entire system. Our model exhibits hysteresis behavior, suggesting that it takes considerable effort to regain trust once it has been broken, emphasizing the self-reinforcing nature of trust at the systemic level. Although simple, the model provides a plausible account of the credit freeze that followed the global financial crisis of 2007/8.

1 Introduction

The sentiment of Trust is vital to the smooth operations of market economies. Its' importance has been recognized by a number of eminent economists, including Dasgupta [4] and Sen [12]. Moreover, it has been argued by Hosking [7] that money is a good yardstick for measuring trust, as it enables us to derive goods and services

Kartik Anand
The Abdus Salam International Centre for Theoretical Physics, Strada Costiera 11, 34014 Trieste, Italy, e-mail: kanand@ictp.it

Prasanna Gai
The Australian National University, Crawford School of Economics and Government, Canberra ACT 0200, Australia,

Matteo Marsili
The Abdus Salam International Centre for Theoretical Physics, Strada Costiera 11, 34014 Trieste, Italy.

M. LiCalzi et al. (eds.), *Progress in Artificial Economics*, Lecture Notes in Economics and Mathematical Systems 645, DOI 10.1007/978-3-642-13947-5_7,

from others, who we do not know, nor have any reason to trust. The explosive growth of credit markets the world over is indicative of trusts' vital role to economic activity. In his treatise on business cycles Schumpeter [11] defined capitalism as "that form of private property economy in which innovations are carried out by means of borrowed money [credit]". With trust enshrined as a social norm, market liquidity is guaranteed and the economy approaches the idealized setting of the General Equilibrium Theory.

However, as witnessed during periodic episodes of financial crises, trust obeys a self-reinforcing dynamic. In this regard, Bagehot [3] noted, 'Credit - the disposition of one man to trust another - is singularly varying [...] after a great calamity, everybody is suspicious of everybody; as soon as the calamity is forgotten, everybody again confides in everybody". Besides the "good" equilibrium where trust is prevailing, there may also be a polar opposite situation where individuals are mis-trusting of each other. Consequently, markets are illiquid and credit is hardly to come by or very costly, making it difficult for individuals to cope with risks and market volatility. As a result, defaults are more frequent, thereby reinforcing the mistrust.

There is no general theory, as far as we are aware of, which describes how and under what conditions trust may evaporate, precipitating an economy from the good state into the bad one. The purpose of the present paper is to contribute to fill this gap, by proposing an agent based model of an (interbank) credit network.

The recent global financial crisis may be viewed as a transition from a good to a bad state, caused by the evaporation of trust. According to Haldane [6], "In essence, events of the past two years can be re-told as a story of the progressive breakdown in trust". The triggers were revelations of losses on United States sub-prime mortgages and other toxic financial assets by banks. An immediate consequence was a freeze in interbank money markets, as banks ceased lending to each other. Mounting funding pressures, in turn, lead to questions about banks' future profitability and, in some cases, viability. Before the crisis, banks required some ten basis points of compensation for making one month loans to each other. By September 2007, that compensation premium had risen to around one hundred basis points. The ensuing collapse of the investment banks Bear Sterns and Lehman Brothers in 2008 led the premium to rise more than thirty-fold from pre-crisis levels. And, despite public sector bailouts of the banking system in the major economies, trust has been slow to return.

While specific details and rigorous economic modeling are necessary to properly understand the financial crisis, there is nonetheless scope to clarify the mechanisms by which trust evaporates in market economies in general, and financial networks in particular. The literature on global games Morris and Shin[9, 10], suggests that financial crises follow a self-fulfilling dynamics, highlighting the importance of *strategic uncertainty* as opposed to *structural uncertainty*. For example, in the case of a bank run, a depositor may consider withdrawing their funds because of doubts over the banks' balance sheet position (structural uncertainty) or because he fears that other depositors are going to withdraw their savings (strategic uncertainty), because they bear similar concerns on the bank or fears on the behavior of other depositors. The coordination game that Morris and Shin [9, 10] focus on is typically a game between multiple lenders involved with a single, risky, counterpart. In the Nash equilibrium of such models, some agents reason that someone else may decide to be cautious and withdraw from the lending arrangement, or at least that someone will reason that someone will withdraw, and so forth. Thus, small seeds of doubt reverberate across all lenders, leading to a wholesale withdrawal of lending and the bankruptcy of the counterpart.

As the recent crisis has made clear, small seeds of doubt about one counterpart can reverberate across the entire global financial system, affecting credit markets in New York as well as Sydney. Agents are likely to be involved in a number of different coordination games against each other. This calls for extending the insights of global games, which pertain to a single financial institution, to the system-level in order to understand wide-spread contagion of skepticism and the collapse of trust.

In this paper, we extend the global game approach to the system level using a model of network growth. Specifically, we show how the arrival of signals about counterparts sows the seeds of distrust and spreads across a large population of agents engaged in credit relationships with each other. Our model, thus, helps shed light on the recent 'freeze' in global interbank lending, in which banks ceased lending to each other, including to well known and long-standing counterparts, once negative signals began to accumulate.

Our results may be summarized as follows: the financial system can converge to a "good" equilibrium in which a dense network of credit relations exists and the risk of a run, and subsequent default, is negligible. But a "bad" equilibrium is also possible – here the credit network is sparse because investors are more skittish and prone to prematurely foreclosing their credit relationships. The transition between the two equilibria is sharp and both states exhibit a degree of resilience; once a crisis tips the system into the sparse state, the restoration of trust requires considerable effort, with model parameters needing to shift well beyond the turning point. And when the system reverts to a good state, it is robust even to deteriorating conditions.

A crucial feature of our model is the rate at which bad news about the creditworthiness of an agent arrives. This, together with the maturity structure of credit contracts, determines the (endogenous) rate of link decay in the network. Intuitively, when bad news arrives an agent may be forced into default by the ensuing foreclosures. This leads to a rearrangement of balance sheets across the financial system – agents who have lent to it lose assets, while agents who borrowed from the defaulter lose liabilities. As a result, there is a possibility that some counterparts may be placed under stress, precipitating further rounds of foreclosures. We discuss the collective properties of the stationary state of these processes.

The focus of the present paper is on computational aspects. So while the main text will mainly discuss the results of numerical simulations, a discussion of the key aspects which allows one to perform fast simulations, is discussed in the appendix.

2 The Model

Consider a population of N agents engaged in bilateral credit relationships with each other. A financial system of this kind can be viewed as a directed network, with nodes representing the agents and outgoing links reflecting loans from one agent to another. To keep matters simple, suppose that all loans take the same unitary nominal value. Hence, the credit network is specified in terms of an adjacency matrix $\mathbf{g} = \{g_{ij}\}$, with $g_{ij} = 1$ if i is lending to j and $g_{ij} = 0$ otherwise.

In our stylized description of the economy, the system evolves according to simple stochastic rules, which we model with Poisson processes occurring at different rates. Links represent short term loans and are used as means to finance long term projects. Projects have a finite life-time, which is modeled by assuming that each loan is terminated and settled by the parties at rate λ. As a consequence, the corresponding link decays ($g_{ij} = 1 \to 0$). At rate ν, which is higher than λ, the lenders to a bank are given the choice of either refinance their loans (i.e. keep the link) or foreclosing them (i.e. remove the link). The ratio ν/λ between the maturity of projects and that of short term loans used to finance it, will be an important parameter of the model. The expected return of each project is R and the interest rate on loans is r.

At any time t, the financial position of agent i is summarized by the assets and liabilities that it holds on its balance sheet. On the asset side, we distinguish between liquid assets – which include holdings of cash b_i and short term loans to other banks $a_i = \sum_{j=1}^{N} g_{ij}$ – and illiquid ones. The latter include the projects financed with short term loans, which amount to $\ell_i = \sum_{j=1}^{N} g_{ji}$. Liabilities, namely the monies owed by agent i to its counterparts, consist of ℓ_i loans taken by bank i (incoming links). Since every outgoing link for one node is an incoming link for another node, it must be that $\langle a \rangle = \langle \ell \rangle$, where the angled brackets refers to the average over all agents. The liability side of the balance sheet includes equity reserves and the regulatory stipulated amount of capital that must be set aside to buffer the bank from negative shocks. If we assume that illiquid assets are only those financed by the ℓ_i loans, then the balance of accounts implies that equity (including regulatory capital) equals $a_i + b_i$.

Each agent i receives new investment opportunities at rate $\gamma + \varepsilon a_i$. When this occurs, agent i solicits funds form a randomly chosen agent j, and we assume that i and j enter in a credit relationship ($g_{ji} : 0 \to 1$, $(\ell_i, a_i) \to (\ell_i + 1, a_i)$ and $(\ell_j, a_j) \to (\ell_j, a_j + 1)$). This implies that, at the onset, loans are made unconditional (or without full knowledge) of a counterpart current positions. This may be seen as a manifestation of trust, *a priori*, between agents. The ε contribution to the rate γ reflects the principle of "excess capacity" [1] : Banks with higher equity levels typically use excess equity to leverage themselves further by borrowing and making new loans. Here we assume that equity can be approximated by a_i, given our discussion above.

In the absence of foreclosures, the two processes of link formation at rate $\gamma + \varepsilon a_i$ and link decay at rate λ (per link), would produce a credit network which is a random graphs [5] with average degree $\gamma/(\lambda - \varepsilon)$. Notice that $\varepsilon < \lambda$ is required to have a stationary network. In this case, if λ is small, i.e. if credit contracts are long lived, these two processes produce a dense credit network.

2.1 Foreclosure Game

Lenders of a agent i have the option to refinance their loans at random Poisson times t_ν, occurring with rate ν. This time may coincide with the time when information on the balance sheet of agent i is disclosed. This includes the number of (liquid) assets a_i and liabilities ℓ_i as well as cash b_i. The cash b_i follows a stochastic evolution with two distinct parts. First, assuming that once interbank loans reach maturity they are repaid at interest rate r, we have that $\dot{b}_{i\,\text{loan}} = \lambda r(a_i - \ell_i)$. Second, the bank may earn returns on the investment it made from loans it borrowed from other banks. This gives the contribution $\dot{b}_{i\,\text{invest}} = \lambda R \ell_i$. Finally, assuming that the cash holdings are consumed at rate χ_i, i.e. $\dot{b}_{i\,\text{cons}} = \chi_i b_i$, we have the stationary state, $b_i \approx \lambda[r(a_i - \ell_i) + R\ell_i]/\chi$. We consequently model b_i as a Poisson variable with mean $\lambda[ra_i + (R-r)\ell_i]/\chi$, which is drawn each time the foreclosure game is played.

At time t_ν, the ℓ_i lenders have a choice of withdrawing their funds (foreclosing) or rolling them over to maturity. Their decision to rollover depends on their opportunity cost of doing so, c_j. If too many agents opt to foreclose, demanding early payment, there is a risk that agent i, is forced into default resulting in poor payoffs for those lenders who decide to rollover. Specifically, if liquid assets $a_i + b_i$ of agent i at time t_ν are not enough to repay all the lenders who decide to foreclose, agent i needs to sell illiquid assets thus causing financial distress or failure. We take the extreme case where if the number κ of lenders who refinance their loans is smaller than the critical threshold $\widehat{\kappa} \equiv (\ell_i - [a_i + b_i])/\ell_i$, bank i will default.

Following [9], we assume that the payoff to player j for opting out is zero. If, however, j rolls over its loan, then the pay off is $1 - c_j$ if $\kappa \geq \widehat{\kappa}$ and $-c_j$, otherwise. We suppose further that the opportunity costs across agents $\{c_j\}$ are determined by a common element and small idiosyncratic elements that introduce small differences in cost around the central tendency. Specifically, $c_j = \theta + s_j$, where θ is the common element in the costs of all agents, and s_j is the idiosyncratic element, which is uniformly distributed over $[-\sigma, \sigma]$. The θ is also uniformly distributed ex ante. Upon observing her own cost, agent j therefore, forms a belief over the fraction of roll-overs, κ.

Following the lines of reasoning proposed by Morris and Shin [9], one may derive a Nash equilibrium solution for this game, which amounts, for each counterpart j, to follow a simple strategy:

$$\begin{cases} \text{if} \quad c_j \leq c^* & \text{rollover} \\ \text{else} & \text{foreclose} \end{cases} \qquad \text{with} \qquad c^* \equiv \frac{a_i + b_i + 1}{\ell_i + 1}. \tag{1}$$

Note that this solution is independent of parameters θ and σ. We now provide the basic intuition for this solution. For a more detailed derivation, we refer the reader to the original paper [9].

We assume that all counterparts are subject to switching strategies, i.e., j will rollover its loan, if its' cost $c_j \leq c^*$, or foreclose, otherwise. Each agent then computes c^* as follows - conjecture that a hypothetical agent h has $c_h = c^*$. We define $\phi = \phi(\theta, \sigma)$ as the probability that no more than $a_i + b_i$ agents have cost greater than c^* – and hence foreclose their loans to i – conditional on $c_h = c^*$. Thus, the expected payoff for h, who is indifferent between rolling over or foreclosing, simplifies to $\phi = c^*$. Finally, ϕ is given by observing that the probability that there are exactly $k = 0, 1, \ldots, \ell_i$ agents with $c_j > c^*$ is $\phi_k = 1/(\ell_i + 1)$. Indeed, *a priori*, ϕ_k cannot depend on k. Therefore $\phi = (a_i + b_i + 1)\phi_k = (a_i + b_i + 1)/(\ell + 1)$, which, combined with the previous result yields Eq. (1).

The independence of this result on σ implies that strategic uncertainty is relevant even in the absence of uncertainty on the payoffs of other players. This is in fact a consequence of modeling θ to have a uniform prior. In what follows, given our emphasis on the collective behavior of the network, we assume $c_j = c$ for all creditors j, irrespective of the counterpart. This allows us to discuss systemic properties in terms of a single parameter c, which is called *cost of miscoordination* in [9].

Summarizing, at rate ν the lenders of each agent i are called to play a foreclosure game. At this time, information on the position (ℓ_i, a_i) of agent i and on their cash holdings b_i are disclosed to all counterparts of i. The analysis described for the coordination game discussed above comes into effect. In particular, all of i's counterparts will foreclose their loans if

$$\ell_i + 1 > c(a_i + b_i + 1)\,. \tag{2}$$

If Eq. (2) is not satisfied, then all of i's counterparts will rollover their loans to the next period. Otherwise, agent i is said to default and is replaced by a new agent with no links, i.e. $(\ell_i, a_i) \to (0, 0)$. Agents, j, who previously borrowed from i will each loose one liability, i.e., $(\ell_j, a_j) \to (\ell_j - 1, a_j)$. Finally, the lenders, k, will loose one asset each, i.e., $(\ell_k, a_k) \to (\ell_k, a_k - 1)$. In reality, lenders may not loose their assets (completely) and borrowers may still be bound to repay their debt. This, however, will occur on longer time scales. As long as we focus on liquid assets and on the short term, the evaporation of links of defaulting banks is a good approximation.

3 Results

In brief, in our model each agent's financial position is specified in terms of its position (ℓ_i, a_i) in the balance sheet plane. The three processes, (i) link addition at rate $\gamma + \varepsilon a_i$ per agent, (ii) link decay at rate λ and (iii) information disclosure, at rate ν per agent, induce a stochastic process in the plane (ℓ, a) of balance sheets. This can be simulated efficiently as discussed in the Appendix. We have run extensive numerical simulations of the model in order to explore it's behavior as key parameters change.

Fig. 1 reports the behavior of the network density $\rho = \langle \ell \rangle$, which is the number of incoming or outgoing links per node, as a function of c. For $\nu = 2$ results for two different population sizes are shown, providing evidence that the density of links in the stationary state is independent of population size. The density ρ (for $\nu = 2$) exhibits a sharp collapse from a dense to a sparse network at $c \approx 0.81$ and a reverse transition at $c \approx 0.6$. For small c, a dense network equilibrium attains, and $\rho = \gamma/(\lambda - \varepsilon)$, indicative of a high level of trust. The news that is released and permeates through the network is encouraging to lenders, who thereby continue to rollover their loans. However, for large c, lenders perceive that the cost of miscoordination from rolling over their loans is high. Thus, doubts concerning the actions of other lenders leads to the collective foreclose of loans, resulting in a sparse financial network. Finally, in an intermediate range of c, we observe the coexistence of both dense and sparse network solutions. This induces hysteresis phenomena: once conditions deteriorate (increasing c) and the system collapses into a sparse network equilibrium, the latter remains stable well beyond the tipping point when conditions improve, and one needs to decrease c to well before the tipping point to regain the dense network solution. This result highlights self-reinforcing nature of trust at the collective level.

The transitions between the two equilibria are documented in detail in Fig. 2, which show time series of the density for values of c close to the transitions. While the transition is sharp and sudden in both cases, a long transient might be necessary before the density collapses from the good to the bad equilibrium, or recovers in the reverse transition. In particular, the recovery of a dense network was found to be quite sensitive to changes in c. For $c = 0.6$ the transition is observed after times which may differ for orders of magnitude, from one realization to the other[1]. The growth of the credit network during the transition is documented in the inset of Fig. 2 (right). This shows how the population of agents rapidly moves away from the unstable region.

Fig. 1 also highlights the non-trivial effects related to the parameter ν. It shows that when ν decreases the stability of the sparse network equilibrium is reduced and, beyond a certain point ($\nu \approx 0.1$) the coexistence region disappears. For values of ν smaller than this, the transition turns into a continuous – although steep – and reversible crossover.

Fig. 3 reports on the dependence of the network density on the (inverse) maturity λ of loans. As λ increases, the density of the two equilibria approach each other,

[1] This behavior is typical of so-called *activated* processes. These are events which are triggered by rare fluctuations, which are large enough to "nucleate" a seed of one equilibrium (the dense network, in our case) in the stationary state of the other equilibrium (the sparse network). Such fluctuations usually have probabilities which are exponentially small in the system's size N. This makes it easier to observe activated processes in smaller systems. Note indeed that the sparse network equilibrium extends a bit further, in the low c region, for $N = 800$ than for $N = 400$ (see Fig. 1).

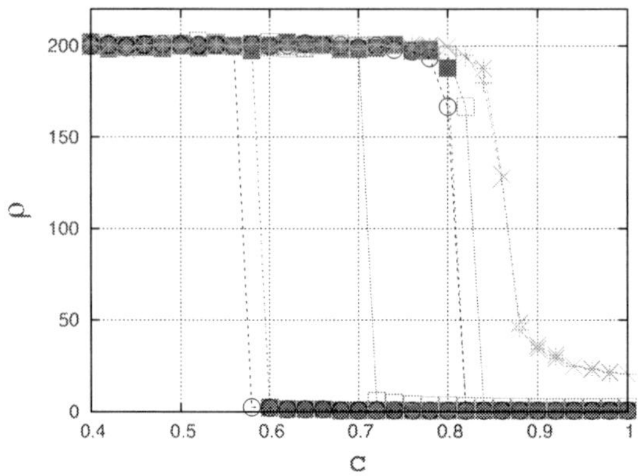

Fig. 1 Average density ρ of the credit network as a function of the cost of miscoordination c and different values of ν and population size N. In particular, $\nu = 2, N = 800$ (open circles), $\nu = 2, N = 400$ (full squares), $\nu = 0.5, N = 400$ (open squares) and $\nu = 0.1, N = 400$ (asterisks). The other parameters are $\lambda = 0.01, \varepsilon = 0.005$, $r/\chi = 0.5$, $(R-r)/\chi = 2.0$ and $\nu = 2$ (open circles and full squares), $\nu = 0.5$ (open squares) and $\nu = 0.1$ (asterisks). Each point is obtained from numerical simulation of the process, for 8000 updates per agent, with averages being taken on the last 500 time-steps per agent. Results are shown both for initial conditions with a sparse and a dense network

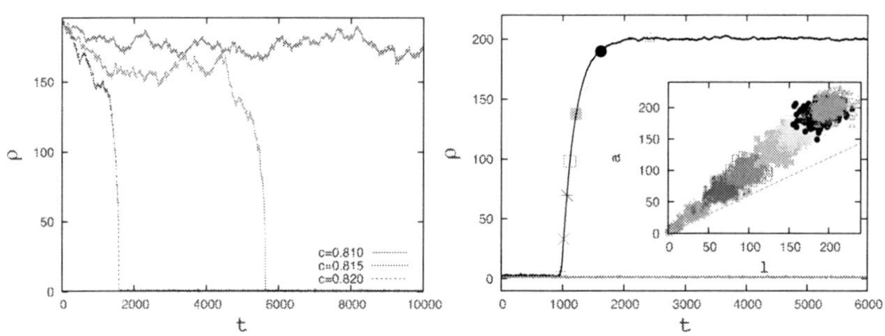

Fig. 2 Time series of the average density ρ of the credit network in a population of $N = 400$ agents close to the crash (left, $c \simeq 0.81$, $\nu = 2$, see Fig. 1) and to recovery (right). For the latter, the transition is observed for $c = 0.6$ (dark line) but not for $c = 0.601$ (light line). The inset shows the evolution of the population in the (ℓ, a) plane at different times (shown with the corresponding symbol in the main plot) during the transition

and for λ large enough ($\lambda \approx 0.2$) the equilibrium morphs continuously from the dense to the sparse network, as c increases. It is worth to remark that the average number of links per agent, in the low density equilibrium, is only dependent on λ. This implies that links decay prior to their natural termination due to the default of agents.

Finally, the dependence on r/χ, R/χ has been found to induce only marginal effects. This can be understood by observing that, independently of r/χ and R/χ, the holding of cash b_i, being proportional to a_i and ℓ_i, is sizable in the dense network

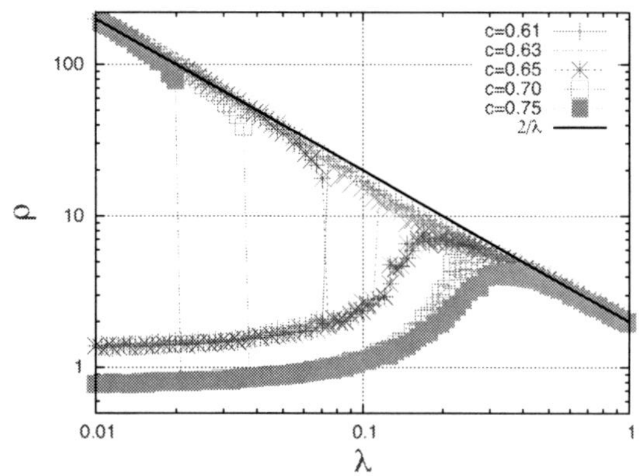

Fig. 3 Average density ρ of the credit network as a function of the rate of link decay λ, for values of c in the coexistence region ($c = 0.61, 0.63, 0.65, 0.70$ and 0.75). Simulations are performed as explained in the caption of Fig. 1 with the same parameters ($N = 400$, $\nu = 2$ and $\varepsilon = \lambda/2$). The full line $2/\lambda$ corresponds to the density of the network without defaults ($\nu = 0$)

equilibrium where $a_i \sim 1/\lambda$, but it is very small in the sparse network equilibrium. Variations of r/χ and R/χ only induce second order effects but are unable to affect the main features of the scenario described above.

4 Discussion

Our model and results highlight elements that were central to the interbank credit freeze that has characterized the recent global financial crisis. First, our model shows how investors may sharply switch to more risk averse behavior, opting for foreclosing their loans and hoarding their liquidity. At the systemic level, this induces drying up of liquidity in the market. Thus, this sharp switch between good and bad states is precipitated by an evaporation of trust between investors.

Second, our model highlights maturity mismatches on balance sheets as a key factor underlying the current crisis as well as the collapse of credit networks in general. Banks financed long-term, illiquid, assets (such as special investment vehicles) by short-term borrowing on the interbank market. This situation corresponds to a large ratio ν/λ in our model, where debts have a long maturity compared to the timescale, $1/\nu$, over which banks refinance their debts by convincing creditors to roll over their loans. It is precisely, and only, in the limit of large ν/λ that a sharp transition such as the one observed in Fig. 1 can occur.

Third, our model sheds light on the nature of public sector intervention during (and since) the crisis. The resumption of normality in the interbank markets required a restoration of trust in the balance sheets of key financial institutions. To facilitate this, central banks cut interest rates to historically low levels, and stepped in guaranteeing for toxic assets on banks' balance sheets, effectively decreasing the cost of miscoordination, c. That these measures had to be protracted for so long time emphasizes the hysteresis entailed in the restoration of trust – a key feature of our model.

An alternative to blanket leverage ratios and liquidity requirements is to target such policies on those financial institutions in the network that are most important [6]. There are interesting parallels here with the literature on attacks on internet-router networks [2]. Extensions of our model along such lines might allow for differential link formation, γ, or preferential linkage where agents in one sub-group prefer to interact with others in the same sub-group. Agent heterogeneity of this kind holds out the possibility of promising new insights into the design of financial stability policy.

The model presented here is simple. In particular, it does not allow for any macroeconomic variability or the exogenous default by a group (or sub-set) of agents. Moreover, it seems likely that the parameters of the model will evolve according to economic conditions and will need to be determined endogenously. Incorporating a richer set of economic interactions into a network setting such as ours is an important step for future research.

5 Appendix: The Algorithm

The population of agents is at each time specified by a configuration $\mathbf{g} = \{g_{ij}\}$ of the credit network, where $g_{ij} = 0$ if i lends to j and $g_{ij} = 0$ otherwise ($g_{ii} = 0$ by convention). Hence $a_i = \sum_j g_{ij}$ is the number of assets and $\ell_j = \sum_i g_{ij}$ is the number of liabilities. The configuration $\mathbf{g}$ is subject to a stochastic dynamics, which is specified in terms of the transition rates $w[\mathbf{g} \to \mathbf{g}']$ and the corresponding master equation [8]. We remind that $w[\mathbf{g} \to \mathbf{g}']dt$ is the probability that a system which is currently in configuration $\mathbf{g}$ will make a transition to configuration $\mathbf{g}'$ in a time interval of duration dt, with dt infinitesimal.

From a computational point of view, the direct way to simulate this process is by setting a small time increment dt and to consider all configurations $\mathbf{g}'$ which can be reached from configuration $\mathbf{g}$, executing each of them with a probability $w[\mathbf{g} \to \mathbf{g}']dt$. This is, however, very inefficient. Indeed dt has to be taken small enough so that at most one of the possible transitions is actually carried out. Since the number of possible transitions is of order N, this amounts to drawing N random numbers to perform at most one transition. In addition to being inefficient, an algorithm based on this procedure is also approximate, because it is not possible to exclude that more than one process will be executed.

An exact method to carry out the simulation is possible, if one makes the following observation. Each process $\mathbf{g} \to \mathbf{g}'$ will occur after a waiting time $\tau_{\mathbf{g}\to\mathbf{g}'}$ which is exponentially distributed, with probability density

$$p_{\mathbf{g}\to\mathbf{g}'}(\tau) = w[\mathbf{g} \to \mathbf{g}']e^{-w[\mathbf{g}\to\mathbf{g}']\tau}. \tag{3}$$

Hence the probability that the next process which occurs is $\mathbf{g} \to \mathbf{g}^*$ can be computed explicitly as

$$P_{\mathbf{g}\to\mathbf{g}^*} = \text{Prob}\{\tau_{\mathbf{g}\to\mathbf{g}^*} \leq \tau_{\mathbf{g}\to\mathbf{g}'}, \forall \mathbf{g}'\} = \frac{w[\mathbf{g}\to\mathbf{g}^*]}{\sum_{\mathbf{g}'} w[\mathbf{g}\to\mathbf{g}']} \tag{4}$$

from the waiting time distributions Eq. (3). Here and in what follows. sums and products on $\mathbf{g}'$ are meant to run on all configurations with $w[\mathbf{g}\to\mathbf{g}'] \neq 0$. In addition, the probability density $p^*(\tau)$ of $\tau_{\mathbf{g}\to\mathbf{g}^*}$ can also be computed easily, taking the derivative of the cumulative density

$$\begin{aligned}\int_\tau^\infty p^*(t)dt &= \text{Prob}\{\tau_{\mathbf{g}\to\mathbf{g}^*} \geq \tau\} \\ &= \text{Prob}\{\tau_{\mathbf{g}\to\mathbf{g}'} \geq \tau, \forall \mathbf{g}'\} = \prod_{\mathbf{g}'} \int_\tau^\infty p_{\mathbf{g}\to\mathbf{g}'}(t)dt = e^{-W_{\mathbf{g}}\tau}\end{aligned}$$

where

$$W_{\mathbf{g}} = \sum_{\mathbf{g}'} w[\mathbf{g}\to\mathbf{g}']. \tag{5}$$

Hence an exact algorithm is possible whereby one selects the configuration $\mathbf{g}^*$ from the distribution in Eq. (4) and then advances time t by an increment dt drawn at random from the distribution density $p^*(dt) = W_{\mathbf{g}}e^{-W_{\mathbf{g}}dt}$.

When transition can be grouped in classes with the same transition rates $w[\mathbf{g}\to\mathbf{g}']$, the algorithm to draw the configuration $\mathbf{g}^*$ can be made much more efficient.

In our case, we can group the transitions in four different classes:

$$w[\mathbf{g}\to\mathbf{g}'] = \begin{cases} \gamma \;\; \mathbf{g}' = (\mathbf{g}_{-ij}, g_{ij}=1) & \text{link addition} \\ \varepsilon a_i \;\; \mathbf{g}' = (\mathbf{g}_{-ij}, g_{ij}=1) & \text{link addition} \\ \lambda a_i \;\; \mathbf{g}' = (\mathbf{g}_{-ij}, g_{ij}=0) & \text{link removal} \\ \nu & \text{foreclosure game} \end{cases}$$

Note that the last process is actually composed of two subprocesses, because the final configuration can be either $\mathbf{g}' = (\mathbf{g}_{-i}, g_{ij} = 0, \forall j)$ if default occurs, or $\mathbf{g}' = \mathbf{g}$ otherwise. Here we used the shorthand notation $\mathbf{g}_{-ij}$ ($\mathbf{g}_{-i}$) for the configuration of the network, excluding link ij (node i).

Our strategy, then, is to draw first the class of the process to be executed, and then draw the process within the class. The latter draw is very efficiently done, as all processes have the same probability, which amount to draw an integer from an uniform distribution (which needs just the draw of a single random number).

In our case,

$$W_{\mathbf{g}} = N(\gamma + \nu) + L_{\mathbf{g}}(\varepsilon + \lambda) \tag{6}$$

where $L_{\mathbf{g}} = \sum_{i,j} g_{ij}$ is the total number of links (assets). So the probability that each of the four processes is executed are

$$p_{\text{class}} = \begin{cases} \frac{\gamma N}{N(\gamma+\nu)+L_{\mathbf{g}}(\varepsilon+\lambda)} & \text{link addition} \\ \frac{\varepsilon L_{\mathbf{g}}}{N(\gamma+\nu)+L_{\mathbf{g}}(\varepsilon+\lambda)} & \text{link addition} \\ \frac{\lambda L_{\mathbf{g}}}{N(\gamma+\nu)+L_{\mathbf{g}}(\varepsilon+\lambda)} & \text{link removal} \\ \frac{\nu N}{N(\gamma+\nu)+L_{\mathbf{g}}(\varepsilon+\lambda)} & \text{foreclosure game} \end{cases}$$

For the first and the fourth process, the second stage involves just the draw of a random node i. In the first case a liability is added to node i. If the foreclosure game is played, the value of b_i will be drawn from the appropriate distribution and default will occur if the instability condition is met. For the second and the third process, the second stage involves choosing a link (asset) at random (an efficient way to do this is to keep an updated list of all links, from which one item is drawn at random). In the λ process, the selected link is removed. In the ε process the selected agent i is identified as the borrower in that credit relationship which corresponds to the link drawn. Then a liability is added to agent i. Finally, time is advanced by an interval dt which is drawn from the exponential distribution with rate given by Eq. (6).

Since the elementary time-step just described involves a number of operation which does not increase with N, this algorithm allows one to perform fast numerical simulations, with N in the order of hundreds or thousands, for times of the order of several thousands updates per agent, in few seconds of a personal computer.

References

1. Adrian T, and Shin H-S (2010) The Changing Nature of Financial Intermediation and the Financial Crisis of 2007-09. Staff Report 439. Federal Reserve Bank of New York.
2. Albert R, Jeong H and Barabasi A-L (2000) Error and attack tolerance of complex networks. *Nature* 406:378-382
3. Bagehot, W., 1873. Lombard Street: A description of the money market. Henry S King and Co, London
4. Dasgupta P (1988) in *In Trust: Making and Breaking Cooperative Relations* (Basil Blackwell, London)
5. Erdös, P., Rényi, A., 1959. On random graphs I. Publicationes Mathematicae Debrecen 6, 290-297.
6. Haldane, A., 2009b. Credit is trust. Retrieved from Bank of England: www.bankofengland.co.uk/publications/ speeches/2009/speech400.pdf.
7. Hosking, G., 2008. The 'credit crunch' and the importance of trust. working paper 77, History and Policy.
8. Gardiner C (2009) *Stochastic Methods: A Handbook for the Natural and Social Sciences* (Springer-Verlag, Berlin), 4th Ed
9. Morris, S., Shin, H. S., 2003. Global Games: Theory and Applications, in: Dewatripont, M., Hansen, L.P., Turnovsky, S. J. (Eds.), Advances in Economics and Econometrics, the Eighth World Congress. Cambridge University Press, Cambridge, pp. 56-114.
10. Morris, S., Shin, H. S., 2008. Financial regulation in a system context, in Elmendorf, D.W., Mankiw, N.G., Summers, L.H (Eds.), Brookings Papers on Economic Activity. Brookings Institute Press, Washington, DC.
11. Schumpeter, J. A., 1939. Business Cycles: A Theoretical, Historical and Statistical Analysis. McGraw-Hill, New York.
12. Sen A (2009) *The New York Review of Books* **56**

Simulations on Correlated Behavior and Social Learning

Andrea Blasco and Paolo Pin

Abstract We consider a population of agents that can choose between two risky technologies: an old one for which they know the expected outcome, and a new one for which they have only a prior. We confront different environments. In the benchmark case agents are isolated and can perform costly experiments to infer the quality of the new technology. In the other cases agents are settled in a network and can observe the outcomes of neighbors. We analyze long–run efficiency of the models. We observe that in expectations the quality of the new technology may be overestimated when there is a network spread of information. This is due to a herding behavior that is efficient only when the new technology is really better than the old one. We also observe that between different network structures there is not a clear dominance.

1 Introduction

We analyze an economy in which agents have to choose between an old technology whose risk is well known, and a new one which not only is risky, but even the probability of success is unknown. They can be thought, as in [4] (inspired by [6]), as farmers that choose between a standard old fertilizer and a new one for which they have only a prior distribution on the quality: henceforth this is the story that we attach to the present exposition. Note that [5] empirically analyzes exactly this situation in a developing country. To be more precise, suppose that each farmer knows that the old fertilizer guarantees a good harvest (payoff normalized to 1) with

Andrea Blasco
Dipartimento di Scienze Economiche, Università di Bologna, Piazza Scaravilli 2, 40127 Bologna, e-mail: andrea.blasco@unibo.it

Paolo Pin
Dipartimento di Economia Politica, Università degli Studi di Siena, Piazza San Francesco 7, 53100 Siena, e-mail: pin3@unisi.it

M. LiCalzi et al. (eds.), *Progress in Artificial Economics*, Lecture Notes in Economics and Mathematical Systems 645, DOI 10.1007/978-3-642-13947-5_8,

probability $\frac{1}{2}$, and a bad one (payoff normalized to 0) with the same probability. On the other hand the new fertilizer will give a good harvest with a probability p that all farmers suppose to be ex–ante uniformly distributed in the interval $[0,1]$. Let us call p the *quality* of the new technology.

The timing of the theoretical model that we consider is discrete and infinite. Farmers, who have full memory and update their beliefs rationally in the Bayesian way, but are infinitely impatient, will repeatedly make their optimal choices year by year. At the beginning of the first year, in expectations, the two fertilizers are equally good, but each farmer would be happy to try out the new technology and gain more information on the real value of p. As time goes on, those farmers that use the old technology and have no information about the new one, will not modify their belief. Those that use the new technology and/or acquire information will get more and more signals on the quality of the new technology and will eventually end up learning which is the best to adopt. However, along any finite step of this process of learning, farmers will find any new information as a valuable good.

We will consider four cases. One in which farmers are isolated and can only perform, on top of past realized harvests, private experiments on the new fertilizer at an exogenously fixed cost c for each of them. Another one in which they are geographically settled on a fixed grid and can, instead of making experiments, observe the last harvest of neighbor farmers who are using the new product, at the same positive cost c for every neighbor they observe. A third one in which they can observe also some farmers which are far away in the grid and not only those in their physical neighborhood. Finally, as fourth case, a completely random directed network. It is clear that in the last three cases we assume a network structure as the one considered in [3, 8], which has been recently generalized by [2]. Our setup is different from that one because agents, when they adopt the old technology and do not pay the cost of observing neighbors, do not get any information. This is why we do not reach uniformity of behavior in the whole economy, as is obtained and studied in the works cited above.

By simply comparing these four cases some interesting questions arise. The possibility of observing neighbors, which can be thought as the sharing of a local public good [4], will have a positive effect on the expected efficiency of our economy, as modeled in [9] and empirically observed in [5], or will it create a herding effect for which all the agents may select the worst technology just because they imitate their neighbors, as happens for consumers in the case analyzed by [7]? Will there be a different outcome if farmers can observe only their grid–neighborhood, compared to the *globalized* cases in which the network structure has long–range links or is even completely random?

In Sect. 2 we introduce the model, while the mathematics underlying the analytical results is in the Appendix. We present the different network structures and report the results of computer–based simulations in Sect. 3. Section 4 concludes.

2 The Model

Let us start from considering the case in which every single agent is isolated. There are two alternative technologies to choose. One gives a reward either 1 with a probability k (that we will fix to $\frac{1}{2}$) or zero. The other gives the good reward with a probability p that is ex–ante distributed on the interval $P \subseteq (0,1)$, with p.d.f. $f_P(p)$. While the realized p is unknown to agents, the distribution is common knowledge. The timing of each step is the following:

- Stage 0: the decision maker does n trials to learn p at a cost $c > 0$ each. The number of possible trials is bounded by N.
- Stage 1: after having observed the outcomes, the project is selected.
- Stage 2: payoffs are realized.

The payoff function is denoted by $U(n,d;p,k)$ and the variable $d = 1 (d = 0)$ represents the agent's decision to pick the risky (safe) asset:

$$U(n,d;p,k) = k + d(p-k) - cn \ .$$

The agent's objective function to maximize is

$$\max_{d \in \{1,0\}} \left\{ \max_{n \in N} E[U(d,n;p,k)] \right\} \ . \tag{1}$$

Fully Bayesian agents will solve the problem backwards [13, 14]. Given $\hat{n}$ observations collected, the agent chooses the optimal policy $d^*(\hat{n})$ and then selects the optimal sample size. The pair (d^*, n^*) denotes the equilibrium strategy profile.

In the Appendix we fully solve analytically this maximization problem when $f_P(p)$ is a Beta function (which generalizes uniform distributions). We use this result, given in (6), to compute numerically the expectations of adopting the new technology if $k = \frac{1}{2}$, $c = 0.004$, $N = 8$ and p ranges between 0 and 0.95. The reasons for which we choose these values for c and N will be explained in next section.

The curve in Fig. 1 shows what is the expected probability that the new technology will be adopted by every single farmer at the time that they stop experimenting (and for each of them the probabilities are i.i.d.). On the x–axis we have the quality p of the new technology, and the vertical dashed line is the threshold below which the new technology is worst than the old one. Because of the assumptions of the model, the fact that farmers may adopt the new technology even when it is worse (i.e. for quality $p < .5$) can be considered in statistical terms as a type II error. Such errors are not a bad result for the long–run efficiency of our economy, because as long as agents use the new technology they get more and more information on its quality p, and will eventually end up dropping it.[1] In terms of long–run efficiency, what should be avoided is the type I error that rejects the new technology even if it

[1] This comes from the fact that all the agents have full information about the old technology. This assumption makes the model analytically tractable.

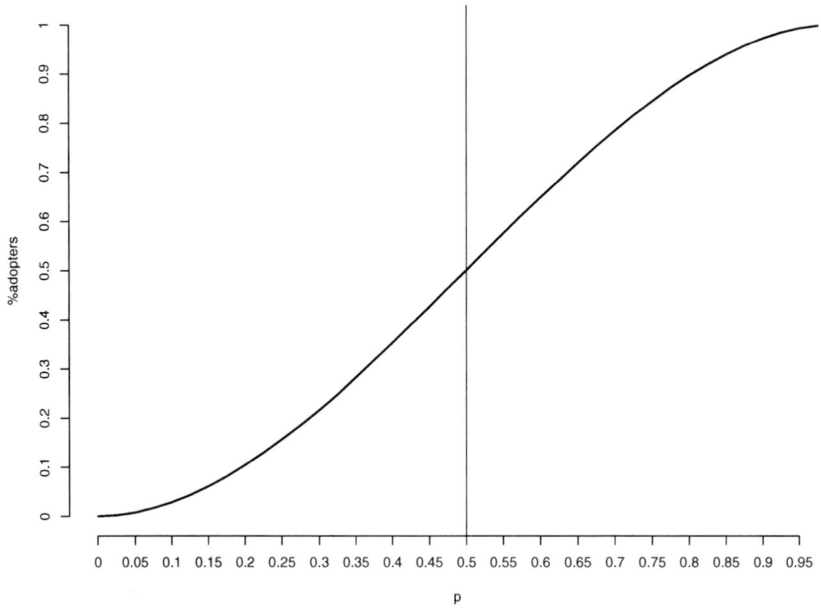

Fig. 1 On the x–axis we have the quality p of the new technology, on the y–axis we have the percentage of agents that are adopting it at the time that they stop acquiring extra information. The curve represents the expected probability of adoption of the new technology. Results are based on equation (6), with parameters $k = \frac{1}{2}$, $c = 0.004$ and $N = 8$

is better than the old one. This may happen frequently when the quality of the new technology is approximately between 0.5 and 0.7.

3 Simulations on Networks

We want to measure and compare the effects of peer imitation and network flow of information. For this reason we consider a new class of environments which have at the individual level the same underlying mathematics, incentives and expected payoffs as the previous one, but is enriched with a new topological structure. We assume now that agents are not isolated but at the same time they cannot make experiments by their own. Instead, they are based in a directed network in which a link from agent i to agent j identifies that i can observe the last harvest (i.e. the outcome of the adopted technology) of agent j.

In particular, we will consider three types of networks. The first one is a square torus grid of 200×200 nodes, where every agent is based on a node and can observe her Moore neighborhood of surrounding 8 nodes. At the end of each time–step she will know wether any one of her 8 neighbors has adopted the old or the new technology. In the latter case she can pay a cost c (which is the same we imposed in previous section for making an experiment) and observe that neighbor. Suppose that agent i has $\ell \leq 8$ neighbors that have used the new technology in the previous time step, then she can, at a cost c per observation, check $n \leq \ell$ of those outcomes (if $n < \ell$ those n are extracted with uniform probabilities). We have chosen the value $c = 0.004$ because, in the simulations described below, we obtain that it makes the threshold $\ell \leq 8$ not more binding than the threshold $N = 8$ that we use in the isolated environment with experiments. We will come back to this point when discussing Fig. 3.

The second class of networks that we consider is the one presented in [12], from the original model of [15]. This *small–world* model is very similar in spirit to the *strong and weak ties* literature in sociology that stems from the work of [10]. Suppose to start from the grid considered above, but now agents keep fixed only their von Neumann neighborhood of four nodes (left, right, up and below), whereas the other four links are cast at random to four among all the other nodes of the grid, with a probability that is proportional to $D^{-\delta}$, where D is the Euclidean distance on the grid between the original node and the candidate node, and δ is a non–negative parameter. The resulting network is a directed one, where node i may observe node j although the opposite may not be true if the two nodes are not von Neumann neighbors. The limiting cases of $\delta \to \infty$ and $\delta = 0$ represent respectively the fixed Moore neighborhood considered above, and the case in which the four new links are casted with uniform probabilities across all the other nodes. Once we have a realization of this random formation process, the obtained network will be exogenously fixed and we will run a round of simulations on it. Figure 2 describes the micro-process underlying this network formation process.

By now we will restrict to the case where $\delta = 0$ and we will discuss more on this point below.

Finally, as a benchmark limiting case, we will consider a completely directed regular random network of out–degree 8, among the 40000 agents that represent

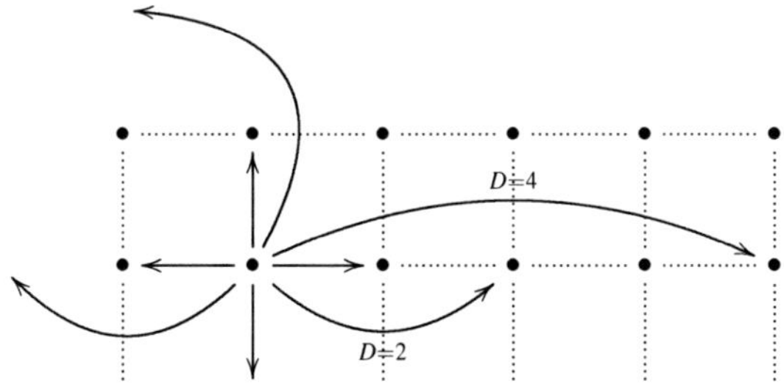

Fig. 2 Graphical description of the Kleinberg [12] model: a node maintains her 4 von Neumann neighbors, and casts 4 links to other nodes with a probability that is proportional to $D^{-\delta}$. D is the geometrical distance to the target and δ is a parameter

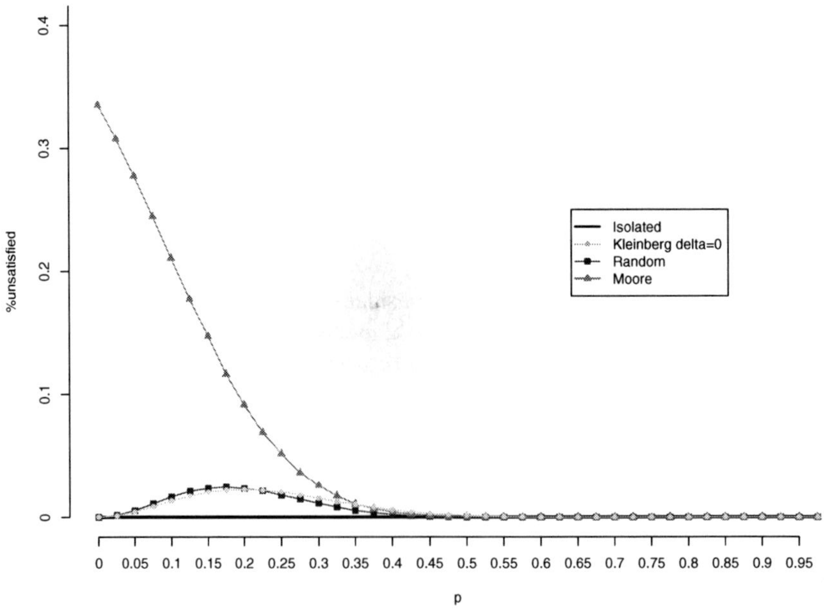

Fig. 3 Results based on 20 iterated simulations with $k = \frac{1}{2}$, $c = 0.004$ and $N = 8$. x–axis shows the quality p. y–axis shows the mean percentage of agents that would have made experiments but are constrained by the threshold of 8 (isolated case) or by the opportunities of making observations (networks)

our population. That is a network in which every node can observe 8 other nodes, but these are drawn at random, with replacement, uniformly and independently, in the whole set of agents.

First of all we compare with the isolated case our three network structures: the Moore neighborhood on the torus grid, the Kleinberg small–world model with $\delta = 0$, and the completely random network. Figure 3 shows the percentage of agents that, under the four cases, are constrained by the threshold imposed to the number of experiments, in the isolated case, or by the number of surrounding neighbors who are still using the new technology (in the remaining network cases).

It should be noted that, as we are mostly interested in type I errors that happen for quality $p > 0.50$, the choice of $c = 0.0004$ and $N = 8$ makes the constrains of the model not binding for the results.

Figures 4 and 5 show the main comparison: the percentage of adopters of the new technology at the time that all agents stop experimenting or observing others' outcomes, leaving the society without any new information. They both represent the result over the same 20 iterations of computer based simulations on different realizations of the networks described by the different models. Everything is repeated

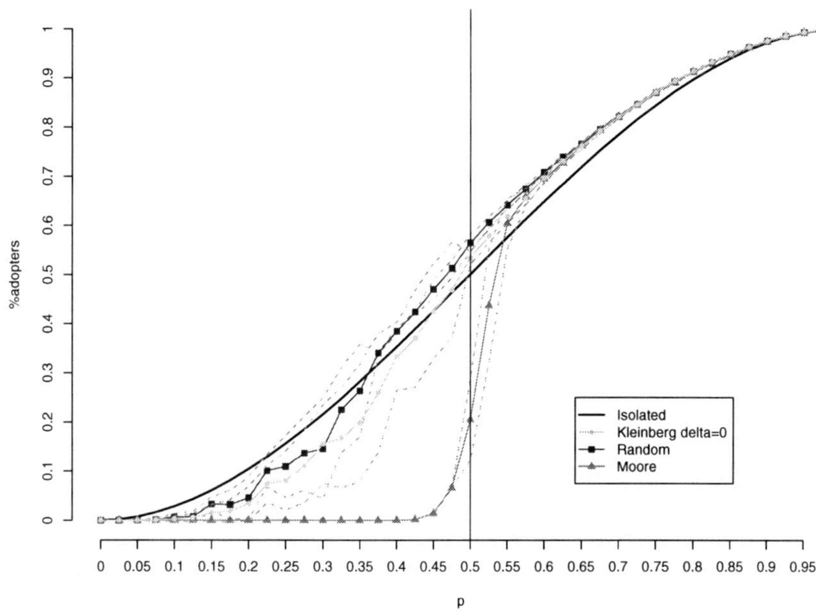

Fig. 4 Results of the simulations with $k = \frac{1}{2}$, $c = 0.004$ and $N = 8$, for the four cases under consideration. On the x–axis is p. On the y–axis are average results, and confidence interval of variances, for the percentage of agents that are adopting the new technology at the time that they stop acquiring extra information

as quality p ranges from 0 to 0.95 with a step of 0.025. Figure 4 considers a larger interval for quality and reports confidence interval of variances, Fig. 5 focuses on a smaller interval and only on mean values. In both figures the black bold curve represents the isolated case from Fig. 1.

Before commenting the result we will further investigate the model of Kleinberg. The theoretical analysis performed in [12] shows that in this model there is a threshold at $\delta = 2$, for the behavior of the agents located in the network, with respect to the flow of information. Kleinberg proves that, for values of $\delta > 2$, the behavior is analogous to the fixed torus grid of Moore neighborhoods; instead, for values of $\delta < 2$, the system behaves as a completely random network. We have not checked this prediction for our specific model, which has a similar structure of information flow, as we rely the analytical results in [12]. We can now generalize and consider essentially three different cases, from Figs. 4 and 5: the benchmark isolated case; the Moore neighborhood on the torus grid; and the random network. The Kleinberg model behaves as the fixed grid for $\delta > 2$, and as the random network for $\delta < 2$.

For values of $p < 0$ we can have type II errors of adopting the new technology when it is actually worse. In such cases the correlated structure of the fixed torus grid

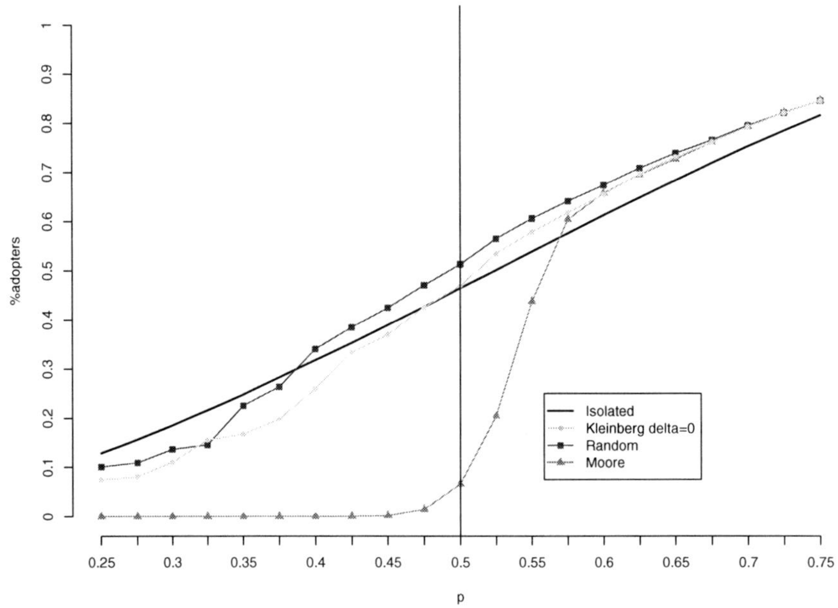

Fig. 5 Results of the same simulations shown in Figure 4. The range for the x–axis is reduced and only mean values are reported on the y–axis

creates a herding behavior that is beneficial to agents in the short–run. However, as the agents in the long–run experience the real quality p of the new technology, this is not so important for the long–run efficiency of our model. For high values of p, more or less above 0.7, all the network structures are slightly more efficient (even in the long–run in this case) than the isolated case: again herding has a positive effect independently on correlations. There is finally an interesting interval when the the new technology is better than the old one but the difference is not strong, i.e. approximately for values of $0.5 < p < 0.70$. In this case the random network still performs better than the isolated case: the long range links provide reinforced information that promotes the spread of the new technology. On the other hand, in the torus grid where information is acquired only in the local neighborhood, and is therefore highly correlated, the network structures performs very badly. This happens because there is a risk of bad initial realizations of the new technology: herding behavior will make the rest and the new technology will be discarded on the basis of few correlated pieces of information.

4 Discussion

We study a model of flow of information concerning the unknown quality p of a new technology that may be adopted instead of a known old one with quality $k = 0.50$. This flow happens on different network structures, and we compare them with a benchmark case in which information is obtained through isolated experiments. The flow of information through a network is characterized by herding effects that may be positive or negative for the long–run efficiency of the model. When the quality of the new technology is actually worse than the old one, then in terms of long–run efficiency our model predicts that, independently on the different assumptions, the agents will end up learning the true quality in any case (i.e. type II errors sooner or later will be discovered and abandoned). When instead the new technology is actually better than the old one, the agents may end up discarding it before they have an accurate enough prediction. What is the effect of a network structure with respect of the risks of a similar type I error to occur?

Our simulations show that the flow of information through the network has a positive effect when the quality of the new technology is really much better than the quality of the old one (a $p > 0.7$ compared to $k = 0.5$). When instead this difference is not so strong, then a random network that avoids the possibility of correlated information is still more efficient. In this latter case we find that instead a fixed grid, in which the flow of information is correlated in the local neighborhood of the agents, performs much worse than the benchmark case in which agents have only the possibility to make isolated experiments on their own.

As the random network always performs as least as well as the grid, for $p > 0.50$ it is clear that an optimal planner should always incentivize a highly uncorrelated network with respect to a geographically constrained one. What if the costs of shifting from one network to the other are high and only partial changes can be implemented? Analyzing the Kleinberg model [12], which is continuous in a single parameter δ, and has as one of the extreme cases the fixed grid, we obtain at the other extremum the same behavior of the random network. Theoretical results in [12] show that this model has actually only two types of behaviors, depending on the value of δ above or below 2. A small value of δ describes the situation in which long–range links, that jump across the fixed geographical proximity, are frequent and break the possibility of correlated behaviors. This point is related to a huge literature in sociology based on the seminal argument of Granovetter [10], who argues exactly that long–range *weak ties*, as he calls them, help enormously in the spread of information across the society.

We hope that the present work could be a starting point for future investigations. An interesting enrichment could be the possibility of letting the agent endogenously create a market and trade their information. Another one could be that of making their utilities depend also on the number of adopters of the new technology, through the kind of externalities analyzed in [11].

Acknowledgements P. P. acknowledges support from the project Prin 2007TKLTSR "Computational markets design and agent–based models of trading behavior".

Appendix

The agent's objective function to maximize is given in equation (1). We model each single experiment as a random variable $X \in \{0,1\}$ with $Pr(X = 1|p) = p$. Let us denote by x_n the vector of realizations out of n trials. Hence, by Bayes' rule, we obtain the posterior probability, given x_n,

$$f_{P|X_1,\ldots,X_n}(p|x_n) = \frac{f_{X_1,\ldots,X_n|P}(x_n|p)f_P(p)}{\int_P f_{X_1,\ldots,X_n|P}(x_n|p)f_P(p)dp} \ . \tag{2}$$

Assuming that $P = (0,1)$ and $f_P(p)$ follows a Beta distribution

$$f_P(p) = \frac{\Gamma(\alpha_0+\beta_0)}{\Gamma(\alpha_0)\Gamma(\beta_0)} p^{\alpha_0-1}(1-p)^{\beta_0-1} \ ,$$

with $\alpha_0 > 0$ and $\beta_0 > 0$ known[2]. We can rewrite the denominator of (2), the probability to observe x_n realizations given the priors, as

$$\int_P f_{X_1,\ldots,X_n|P}(x_n|p)f_P(p)dp = \int_0^1 p^{\sum x_i}(1-p)^{n-\sum x_i} \frac{\Gamma(\alpha_0+\beta_0)}{\Gamma(\alpha_0)\Gamma(\beta_0)} p^{\alpha_0-1}(1-p)^{\beta_0-1} \ .$$

Hence, by denoting $\sum x_i \equiv y$, we obtain an explicit form for the posterior distribution, that is the p.d.f of p after having observed the data:

$$f_{P|X_1,\ldots,X_n}(p|x_n) = \frac{p^{\alpha_0+y-1}(1-p)^{\beta_0+n-y-1}}{\int_0^1 p^{\alpha_0+y-1}(1-p)^{\beta_0+n-y-1}dp} \ . \tag{3}$$

Following [1], we can simplify (3) to

$$f_{P|X_1,\ldots,X_n}(p|x_n) = \frac{\Gamma(\alpha+\beta)}{\Gamma(\alpha)\Gamma(\beta)} p^{\alpha-1}(1-p)^{\beta-1} \ . \tag{4}$$

As a result, the posteriors of p follow the Beta distribution as well but with reassigned constants. Relabeled, here, as $\alpha = \alpha_0 + y$ and $\beta = \beta_0 + n - y$. The conditional mean of p is

$$E[p|X_1,\ldots,X_n] = \frac{\alpha}{\beta+\alpha} = \frac{\alpha_0+y-1}{\alpha_0+\beta_0+n-1}$$

and the variance,

$$VAR[p|X_1,\ldots,X_n] = \frac{\alpha\beta}{(\alpha+\beta)(\alpha+\beta+1)}$$

This formulation is convenient since now we can set $\alpha_0 = y_0 + 1$ and $\beta_0 = n_0 - y_0 + 1$ and interpret the initial prior distribution as the posteriors of p with y_0 positive realizations being observed out of n_0 trials. It generalizes our model to the case in which

[2] It encompasses the case of the Uniform distribution for $\alpha_0 = \beta_0 = 1$

an agent has already observed some data and decides how much new observations to acquire:

$$f_{P|X_1,\ldots,X_n}(p|x_n) = \frac{\Gamma(n_0+1+n)}{\Gamma(y_0+1+y)\Gamma(n_0-y_0+1+n-y)} p^{y_0+y}(1-p)^{n_0-y_0+n-y} \ .$$

Finally, for a given sample size $\hat{n}$, the objective function is easily rewritten as

$$\max_d \ E[U(d;\hat{n},p,k)] = k + d\left(\frac{y_0+y}{1+n_0+\hat{n}} - k\right) \ , \tag{5}$$

leading to the following optimal policy:

$$d^* = 1 \text{ if } \frac{y_0+y}{1+n_0+\hat{n}} \geq k \quad ; \quad d^* = 0 \text{ otherwise} \ .$$

It is equivalent to define a threshold

$$\bar{x} = k(n+n_0+2) - x_0 - 1$$

to obtain

$$\max_{n\in N} U = \frac{\binom{n_0}{x_0}(n_0+1)}{(n+n_0+1)}\left[\sum_{x=0}^{\bar{x}-1} k\frac{\binom{n}{x}}{\binom{n+n_0}{x+x_0}} + \sum_{x=\bar{x}}^{n} \frac{x_0+x+1}{n_0+n+2}\frac{\binom{n}{x}}{\binom{n+n_0}{x+x_0}}\right] - cn$$

Given this solution we are able to solve also the sample size's problem:

$$\begin{aligned}
\max_{n\in N} U &= \sum_{y=0}^{n} E\left[U(d^*(n),n;p,k)|X_1,\ldots,X_n\right] f_{X_1,\ldots,X_n}(x_n) - cn \\
&= \frac{\Gamma(\alpha_0+\beta_0)}{\Gamma(\alpha_0)\Gamma(\beta_0)} \sum_{y=0}^{n} \max\left[\frac{\alpha}{\alpha+\beta}, k\right] \int_0^1 p^{\alpha-1}(1-p)^{\beta-1} dp - cn \\
&= \frac{\Gamma(\alpha_0+\beta_0)}{\Gamma(\alpha_0)\Gamma(\beta_0)} \sum_{y=0}^{n} \max\left[\frac{\alpha_0+y}{\alpha_0+\beta_0+n}, k\right] \frac{\Gamma(\alpha_0+y)\Gamma(\beta_0+n-y)}{\Gamma(\alpha_0+\beta_0+n)} - cn \ .
\end{aligned} \tag{6}$$

We base on equation (6) the numerical computations of the expected behaviors of our model.

References

1. Abramowitz M, Stegun I A (1965) Handbook of mathematical functions. New York
2. Acemoglu D, Dahlehz M A, Lobel I, Ozdaglar A (2010) Bayesian Learning in Social Networks. Mimeo
3. Bala V, Goyal S (1998) Learning from Neighbours.Review of Economic Studies **65**: 595–621
4. Bramoullé Y, Kranton R (2007) Public goods in networks. Journal of Economic Theory **135**: 478–494
5. Conley T G, Udry C R (2010) Learning about a New Technology: Pineapple in Ghana. American Economic Review **100**(1): 35–69

6. Foster A D, Rosenzweig M R (1995) Learning by Doing and Learning from Others: Human Capital and Technical Change in Agriculture. The Journal of Political Economy **103**(6): 1176–1209
7. Galeotti A (2010) Talking, Searching and Pricing. International Economic Review. Forthcoming
8. Gale D, Kariv S (2003) Bayesian learning in social networks. Games and Economic Behavior **45**: 329–346
9. Golub B, Jackson M (2010) Naive Learning in Social Networks: Convergence, Influence and the Wisdom of Crowds. American Economic Journal: Microeconomics **2**(1): 112–149
10. Granovetter M (1973) The Strength of Weak Ties. American Journal of Sociology **78**: 1360–1380
11. Kats M L, Shapiro C (1985) Network Externalities, Competition, and Compatibility. American Economic Review **75**(3): 424–440
12. Kleinberg J (2000) The small-world phenomenon: An algorithmic perspective. Procedings of the 32nd ACM Symposium on Theory of Computing
13. Lam Y, Li K–H, Ip W–C, Wong H (2006) Sequential variable sampling plan for normal distribution. European Journal of Operational Research **172**: 127–145
14. Lindley D V (1997) The choice of sample size. In: The Statistician: 129–138. Blackwell Publishers
15. Strogatz S, Watts D J (1998) Collective dynamics of small-world networks. Nature **393**: 440–442

Technology Shocks and Trade in a Network

How business cycles emerge from the interaction of autonomous agents

Davoud Taghawi-Nejad

Abstract In this paper we show how business cycles can emerge from the interaction of autonomous agents. We devised an agent-based computational microeconomics model of agents who trade in a network of trading partners. We assume that agents who observe decreased profits change their trading partners. At fixed intervals a new production technology becomes available to a single agent. If an agent introduces a new technology he changes his trading pattern and some of his trade partners can have a decrease in profits. The agents who have lower profits start changing trading partners. The change in the trading network can lead to lower production and decreased profits of other agents. Agents with decreased profits also start changing trade partners. In short, the technology shock triggers a snowball effect of agents changing their trading partners; the GDP decreases. When agents find new trading partners and regain their profits the GDP increases. A business cycle emerges.

1 Introduction

In this paper I show how the technological innovation of a single agent can lead to a temporary decrease in GDP and hence to business cycles. The business cycles are generated by the actions and interactions of autonomous agents producing and trading in an endogenously changing network. I do neither assume nor explicitly model business-cycles. With this approach I follow the research agenda outlined by Robert Axtell (2006) and Joshua Epstein (1999) in "Multi-Agent Systems Macro: A Prospectus" and "Agent-Based Computational Models And Generative Social Science", respectively.

In order to devise an agent-based model that leads to business cycles, I abandon the assumption of a central market and replace it with the assumption that firms

Davoud Taghawi-Nejad
Department of Economics "S. Cognetti de Martiis", Via Po, 53. I-10124 Torino, Italy
e-mail: Davoud@Taghawi-Nejad.de

M. LiCalzi et al. (eds.), *Progress in Artificial Economics*, Lecture Notes in Economics and Mathematical Systems 645, DOI 10.1007/978-3-642-13947-5_9,

have a network of suppliers and purchasers. In this respect the model is similar to Wilhite's (2001) paper on "Bilateral Trade and 'Small-World' Networks". However, contrary to Wilhite's model the network in this paper is not fixed, but changes endogenously.

For example if a firm which has several suppliers loses one of them it will produce less until it finds a new supplier. That means that a change in the network can decrease output. Introducing a mechanism that leads to an initially increasing and then decreasing number of firms that change partners, would lead to business cycles.

The mechanism is the following: firms change partners if and only if their profits decrease. The justification for this is that changing suppliers or purchasers is costly. If firms change their trade partners, some of their trade partners and their indirect trade partners might have a decrease in profits. The negatively affected firms start changing their trading partners, too. A snowball effect arises: the number of firms that experience a decrease in profits and change trading partners increases.

When firms regain their profit, because they found new trading partners, they stop changing their trading partners. At this point the number of firms that change partners decreases and consequently the number of firms that have decreased profits decreases. We have a cyclical pattern in the GDP.

A technological shock could trigger this cycle: when a particular firm introduces a new technology, their demand for and supply of production factors changes. This can affect the profits of their suppliers and purchasers. The avalanche is launched. The technological shock triggered a business-cycle in the GDP.

2 The Model

The model's three components are four different types of agents, four different factors of production, which can be traded and then used for production of the third ingredient, the final good. The four kinds of agents differ only in the fact that they start every round with an endowment of a different factor of production. This makes trade necessary. The final good is the medium of exchange in the economy. An agent's wealth is the amount of the final good he possess. Agents trade the factors of production only with agents they have connections with.

The model is implemented as an agent-based computer program. The source code is available at `http://taghawi-nejad.de/index.php?id=69`.

2.1 Agents

Each type of agent is *ex ante* homogeneous. The only difference among agent types is that they produce different factors of production. The agents are neo-classical in the sense that they optimize in a marginalist manner. Contrary to standard neo-classical models, the agents do not trade in a central market and have incomplete knowledge about the future. They do not know how the network evolves and which new technologies will be developed in the future. Agents maximize their wealth or equivalently maximize their profits, the net gains from trade and production minus the costs of maintaining trade relationships.

The type of an agent is a 4-tuple (a,b,c,d) which denotes the endowments for the four factors of production each agent gets at the beginning of each round. Each agent has four tradable factors of production(a_i,b_i,c_i,d_i), income I_i, wealth W_i and a list of trading partners F_i. Furthermore, the production technology of each agent is represented by a set of variables, which are the exponents of the Cobb-Douglas technology.

$$y = 100 \cdot a^{\alpha} \cdot b^{\beta} \cdot c^{\gamma} \cdot d^{\delta} \qquad \alpha+\beta+\gamma+\delta = 1$$

Different exponents represent different production technologies.

2.2 Rounds and Agents' Actions

The simulation runs over several rounds each consisting of 7 sub-functions. In every sub-round each agent, in sequential order, executes one rule; the only exception is step 2 of stage 1, where rule 2 (Trade) gets executed 100 times alternating a sequential and reversed order. In stage 3, the agents execute rule 6 and 7 respectively, only when rule 5 says so. The sub-functions are listed below; the rules will be exposed in the remainder of this section.

In the first stage, the network in which the agents operate is given. Therefore the agents behave in this part like standard neoclassical agents. In the second stage agents redesign their network. In the third stage they change technology.

In each of the steps in the second and the third stage, agents have to take three decisions: whether to add a trading partner, whether to delete a trading partner and whether to implement a new technology. The agents decide between two options by calculating the respective profits. In order to calculate the profits they perform a virtual computation (in their "head") twice through stage 1, once with and once without the considered change.

Round

Stage 1

1. Replenishment of the factor of production (Rule 1)
2. Trade with trading partners (Rule 2 repeated 100 times in alternating order)
3. Production of the final good (Rule 3)
4. Paying the cost for trading partners (Rule 4)

Stage 2

1. Terminate connections with unprofitable trading partners (Rule 5 $\rightarrow$ Rule 6)

2. Establish connections with new trading partners (Rule 5 → Rule 7)

Stage 3

1. Change technology. (Rule 8)

Replenishment of the Factors of Production

Every agent replenishes his endowment of the factor of production. 50 units of the factor of production that corresponds to his type and one of every other factor of production.

Rule 1 *Replenish your endowment of the factor of production of the type that corresponds to your type to 50 and to 1 for every other type.*

Trade

An agent trades exclusively with his trading partners. They use the final good as the medium of exchange. The trading agent asks all his trading partners for the marginal productivity of the last factor of production of each type ($MP_{(-1)seller}$). The agent then buys the most profitable factor of production; that is, the one with the biggest difference between his marginal productivity (MP_{buyer}) and the price.

Definition 1 $price = \frac{MP_{buyer} - MP_{(-1)seller}}{2}$

Agents trade 100 times, twice as much as they produce the factors of production. This allows them to engage in buying and reselling. The effect of this is that agents who are not directly connected influence each other.[1]

Rule 2

1. Ask all trading partners for the marginal productivity of the last factor of production of each type.
2. Buy the factor of production that gives the highest $MP_{buyer} - price$ from the cheapest seller, except if $MP_{buyer} < price$ for all factors of production.

Final Good Production

Rule 3 *Use all your factors of production for the production of the final good, using the Cobb-Douglas function below, with your set of coefficients.*

$$y = 100 \cdot a^{\alpha} \cdot b^{\beta} \cdot c^{\gamma} \cdot d^{\delta} \qquad \alpha + \beta + \gamma + \delta = 1$$

[1] To minimize systematic biases the agents trade once in sequence of their ID and then once in reverse order.

Cost for Trading Partners

We assume that it is costly to maintain a social network. The costs increase linearly with the number of connections an agent has.

Assumption 1 *Maintaining a trading partner costs TradingPartnerCost per round.*

Here, TradingPartnerCost is a model parameter.

Rule 4 *For each trading partner, deduct TradingPartnerCost from your wealth.*

Decision to Change Trading Partners

Searching for new agents is costly, and the price per round is SearchCost. As the gains of a search are uncertain and unquantifiable, they cannot be compared with the price of the search. Therefore, we introduce a decision rule: terminate connections to partners and search for new partners only when your profits decreased or when you were trying to change partners the round before and your profits did not recover.

Agents expect that the search leads to at least the same income they had the round before they started searching. After the first round they decrease their expectations by a given percentage (ρ) each round.

Assumption 2 *Searching for new partners costs SearchCost per round.*

Here, SearchCost is a model parameter.

Condition 1 *The agent has no trading partner.*

Condition 2 $I_t < I_{t-1}$.

Condition 3 *The agent was changing partners the round before and* $I_t < I_s \cdot \rho^r$.

Here, I_s is the income of the agent at the time he started searching; ρ is a model parameter between 0 and 1; and r is the number of rounds the agent is searching.

Rule 5 *If and only if condition 1, 2 or 3 is met, pay SearchCost, apply rule 6 to existing trading partners and rule 7 to new trading partners in the ordered list* M_j.

Terminating Connections with Unprofitable Trading Partners

If an agent decides to change trading partners he starts by terminating unprofitable connections.

${}_jI_{-i}$ is the income of an agent j under the assumption that the agent i gets deleted from the network of trading partners.

${}_jI$ is the income of agent j under the assumption that the agent i is not deleted.

If the following condition holds a connection is unprofitable:

Condition 4

$$
{}_jI_{-i} - TradingPartnerCost \cdot (\#TradingPartner_j - 1) > {}_jI - TradingPartnerCost \cdot \#TradingPartner_j
$$

#TradingPartner is the number of trading partners an agent has.

Rule 6 *For all a_i in F_j, if condition 4 holds than delete agent a_i from list F_j and a_j from list F_i.*

Meeting New Agents

First, nature decides according to a geometrical distribution how many new agents the agent i is going to meet. The searching agent can meet any of the agents in the model, but is more likely to meet agents in his proximity. Nature creates an ordered list M_i of agents that agent i is going to meet.

Agents a_j are ranked by m_{ij}. The ordered list M_i contains the first x agents, where x is a geometrically distributed random variable with $p = 0.5$.

$$
m_{ij} = \varphi \left(\frac{1}{d_{ij}} \right)^2 p_{ij} + (1 - \varphi) p_{ij} \qquad 0 < \varphi < 1 \tag{1}
$$

Here, d_{ij} is the normalized shortest distance between i and j in the network; and p_{ij} is a uniformly distributed random number between 0 and 1.

To interpret the ranking criteria, let's assume $\phi = 1$. The distance multiplier $(d_{ij})^2$ is higher the closer the agents i and j are in the social network. Thus closer agents are more likely to be ranked high. If $\phi = 0$ then the ranking of each agent is completely random. For intermediate ϕ, the resulting network has small world network properties; it is random but localized.

Deciding Whether to Add the New Agents

An agent meets the agents that nature passes him in list M_j. If two agents meet then both agents decide whether it is profitable to establish a trade relation with each other. They get connected only if both of them agree. An additional connection is costly.

Let ${}_jI_{+i}$ be the income of agent j under the assumption that a new agent i gets added to the network of his trading partners. Let ${}_jI$ be the income of agent j under the assumption that the agent is not added.

Condition 5

$$
{}_jI_{+i} - TradingPartnerCost \cdot (\#TradingPartner_j + 1) > \\ {}_jI - TradingPartnerCost \cdot \#TradingPartner_j
$$

Condition 6

$$
{}_iI_{+j} - TradingPartnerCost \cdot (\#TradingPartner_j + 1) > \\ {}_iI - TradingPartnerCost \cdot \#TradingPartner_i
$$

Rule 7 *For all a_i in M_j in the order of M_j, if condition 5 and condition 6 hold, then add agent a_i to list F_j and a_j to list F_i.*

Research

Each agent runs a search every ς rounds, where ς is one of the model parameters. [2] Researching means randomly drawing a new technology. Drawing a new set of the variables $(\alpha, \beta, \gamma, \delta)$ which then implies a new candidate Cobb-Douglas production function, with new exponents.

The new Cobb-Douglas exponents are a linear combination between the old Cobb-Douglas exponents and a univariate random variable on $[0,1]$. Then the exponents are divided by the sum of all exponents to ensure that the Cobb-Douglas function exhibits constant returns to scale. If the research weight parameter $rw = 1$, the research is completely random. The Cobb-Douglas exponents are bounded above by 0.5.

$$
\begin{aligned}
\alpha' &= (1 - rw) \cdot \alpha + rw \cdot a \\
\beta' &= (1 - rw) \cdot \beta + rw \cdot b \\
\gamma' &= (1 - rw) \cdot \gamma + rw \cdot c \\
\delta' &= (1 - rw) \cdot \delta + rw \cdot d
\end{aligned}
$$

[2] The parameter ς establishes the frequency with which the search is done. The exact round at which an agent does a search is co-determined by the agents' ID. For example, if the research frequency is 50, then agent 1 searches in rounds 1, 51,101,151..., while agent 5 searches in rounds 5, 55, 105,155 ...

$$\alpha_{new} = \frac{\alpha'}{(\alpha' + \beta' + \gamma' + \delta')}$$

$$\beta_{new} = \frac{\beta'}{(\alpha' + \beta' + \gamma' + \delta')}$$

$$\gamma_{new} = \frac{\gamma'}{(\alpha' + \beta' + \gamma' + \delta')}$$

$$\delta_{new} = \frac{\delta'}{(\alpha' + \beta' + \gamma' + \delta')}$$

Here, the research weight rw is a model parameter and a, b, c, d are univariate random variables between 0 and 1.

$$\alpha_{new} < 0.5, \beta_{new} < 0.5, \gamma_{new} < 0.5, \delta_{new} < 0.5$$

Technology Implementation Decision

When a new technology is implemented the machines need to be changed and people have to be trained while managerial capacity is bounded. Therefore, we assume that the agents make no profits in the round after they search.

Assumption 3 *When an agent implements a new technology the price is next round's profits.*

The agents calculate the future profits with and without the new technology and subtract the cost of change. If profitable, they implement the new technology.

If the following equation holds for the new technology, it is profitable and therefore it gets implemented:

Condition 7

$$I_{new} - I_{old} > \frac{I}{\iota}$$

where I_{new} is the income under the assumption that the new technology $(\alpha_{new}, \beta_{new}, \gamma_{new}, \delta_{new})$ is implemented; and I_{od} is the income under the assumption that the current technology remains in place; ι is a model parameter, the time in which an agent wants to recuperate his investment

Rule 8 *If condition 7 holds then* $\alpha = \alpha_{new}, \beta = \beta_{new}, \gamma = \gamma_{new}, \delta = \delta_{new}$.

If the new technology is profitable within ι rounds, then the new technology is implemented.

3 Analysis of the Agent-Based Model

In order to draw macro-economic conclusions from the model, we run simulations with different parameterisations. Figure 1 shows a specific but otherwise representative run. One can see that the GDP is increasing at a diminishing rate.

Result 1 *If agents trade exclusively with their trading partners, produce and change their trading partners when unprofitable and innovate their technology according to rules 1–8, then the collective output (GDP) grows at a diminishing rate.*

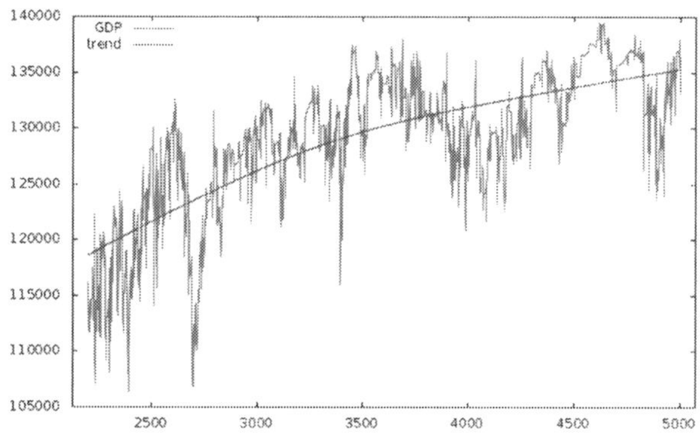

Fig. 1 GDP of the simulation with specification 1

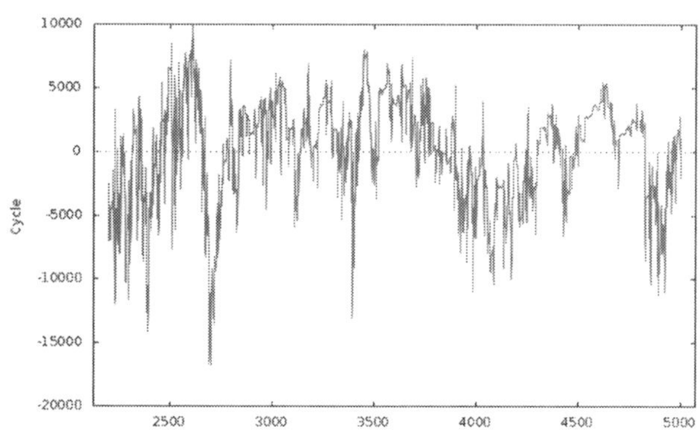

Fig. 2 Detrended GDP of the simulation with specification 1

Figure 2 shows the GDP, detrended with the Hodrick-Prescott filter. A visual inspection of the detrended GDP indicates that the GDP follows a cyclical pattern. If the process was a random walk, we could falsely conclude cyclicality. Therefore we have to rule out that the process is a random walk by a statistical test. We can reject that the GDP follows a random walk at a 0.1 percent level[3]. Secondly, because the network shape and the technological changes rely on a random process, the exact form of the graph in Fig. 1 depends on the simulation. By inspecting the autocorrelogram, we can gain insights about the underlying process.

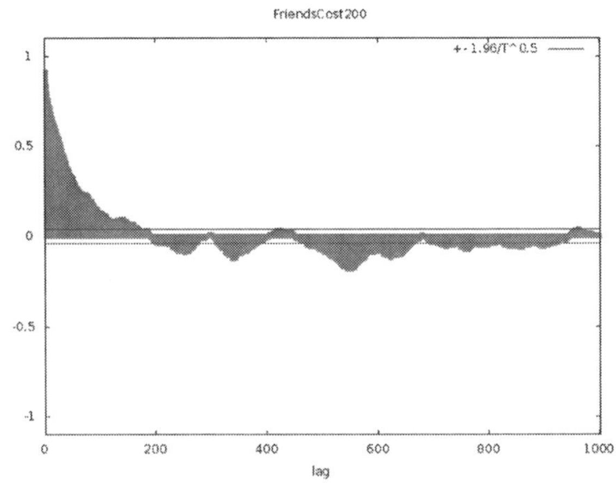

Fig. 3 The ACF of the simulation with parametrization 1

The autocorrelogram of the detrended GDP in Fig. 3 shows a cyclical pattern.

Result 2 *If agents trade exclusively with their trading partners, produce and change their trading partners when unprofitable and innovate their technology according to rules 1–8, then the collective output (GDP) is cyclical.*

For the brevity of the exposition I will only discuss Result 2 here.

3.1 Business Cycles

Business cycles are the consequence of a snowball effect of agents changing partners. The snowball effect is triggered by technological innovation. A technological innovation can decrease the income of the partners of the innovating agent. Decreased income triggers them to change their partners. When agents change partners, their existing partners can have a decrease in income. This triggers them to change their partners too; more and more agents change partners. Because agents who lose a partner or whose partner trades less with them, produce less than before, the increasing number of changing agents manifests itself in a decreased GDP.

[3] The statistical test is reproduced in the statistical appendix of the working paper. `http://www.taghawi-nejad.de/assets/Resources/thesis.pdf`

Agents finding new partners increase their profits. When the agents eventually connect to new trading partners and their income recovers, they stop changing partners. The GDP recovers.

There are two effects at work: first, technological innovation triggers trade partners of the innovating agent to change partners. Second, agents who have decreased profits because their trade partners changed trade partners, start changing their partners. The left image in Fig. 4 shows what happens when an agent innovates. The middle image shows the resulting changes in the network. This picture shows that the change in the network decreases the profits of some agents. The right picture shows changes in the network that result from the second wave of changing.

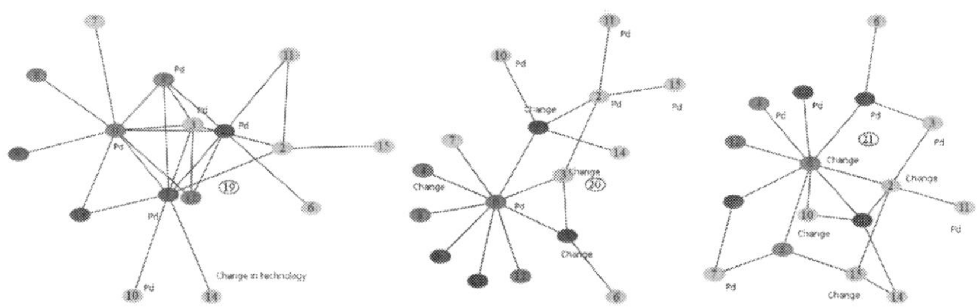

Fig. 4 The left image shows the effects of an innovating agent (Pd = Profits decrease). The middle image shows the resulting changes in the network and how these decrease the profits of other agents. The right picture shows changes in the network that result from the second wave of changes

4 Growth in a Non-Growth Model

In this model there is growth, in spite of the constant returns to scale assumption and the absence of a capital accumulation. The growth has two sources, the increasingly better adaptation of the production function and the optimization of the network. A technology is better adapted when it leads, given the network and other agents price functions, to a higher income than another technology. Different networks can lead to a different GDP as factors of production can be allocated differently. A network, given the technology, is more optimized when it leads to a higher GDP.

5 Conclusion

We have devised a micro-economic model of agents who trade with a limited number of trading partners. Agents who experience decreased profits change their trading partners. At fixed intervals a new production technology becomes available to an agent, and this is an exogenous technology shock. The technological shock can trigger a snowball effect of agents changing their trading partners. Agents whose trading partners change their trade connections produce less. As a result the total output shows a cyclical pattern.

With result 1 we show that the model has an increasing GDP, in spite of the absence of capital accumulation and constant returns to scale. With result 2 we show that the model can lead to business cycles. In my working paper I show that the result is robust to a change in parametrization. I conclude therefore that new technologies can decrease the GDP temporarily and increase it in the long-run. Stated differently, technological innovation can induce business-cycles. In contrast to the mainstream real-business cycle theory, we have not modeled business cycles, but shown how business cycles can be created from the interaction of autonomous agents.

References

1. Axtell, R. (2006) Multi-Agent Systems Macro: A Prospectus. In Colander, D. (ed.), Prepared for Post-Walrasian Macroeconomics, Cambridge University Press, Cambridge
2. Carvalho, V. (2007) Aggregate Fluctuations and The Network Structure of Intersectoral Trade. Available at `http://www.crei.cat/people/carvalho/papers.html`
3. Epstein, J.M (1999) Agent-Based Computational Models and Generative Social Science. Complexity **4**(5):41–60
4. Wilhite, A. (2001) Bilateral Trade and Small-World Networks. Computational Economics **18**(1):49–64

Part IV
Management

The (Beneficial) Role of Informational Imperfections in Enhancing Organisational Performance

Friederike Wall

Abstract The paper analyses the effects of imperfect information on organisational performance under the regime of alternative organisational settings. The analysis is based on an agent-based model which is an extended variant of the NK model. In the simulations, fitness measurements are distorted with imperfections according to information asymmetries that are related to differentiation and delegation of decision-making. The results indicate that the effects of informational imperfections on organisational performance subtly interfere with coordination mode, incentives and intra-organisational interactions. The results might throw some new light on imperfect information as in some organisational settings rather insignificant performance losses compared to perfectly informed decision-makers occur, and in some settings imperfect information turn out to be beneficial.

1 Introduction

Limited information processing capabilities of individuals and imperfect information in decision-making have been analysed in numerous organisation theories and with respect to various elements of organisational design that affect the overall structure of an organisation. For example, in their seminal work Simon and March regard differentiation and delegation as means to reduce uncertainty and complexity in decision-making and, thus, to reduce information needs of decision-makers [14, 21]. Contingency theories treat information processing needs resulting from complexity and uncertainty as a situational determinant for organisational design and information processing capacities as means to adjust an organisation to the situation, e.g. [7, 8, 24]. Team theory points out that by differentiation and delegation

Friederike Wall
Department of Controlling and Strategic Management, School of Management and Economics, University of Klagenfurt, Austria
e-mail: friederike.wall@uni-klu.ac.at

M. LiCalzi et al. (eds.), *Progress in Artificial Economics*, Lecture Notes in Economics and Mathematical Systems 645, DOI 10.1007/978-3-642-13947-5_10,

decisions can be made on refined information structures [15]. Following agency theory the principal delegates a decision to an agent to benefit from the agent's information (capabilities); the principal, however, accepts that information asymmetries and conflicting interests may cause agency problems [6, 12].

Design elements that mainly focus on the behaviour of individual decision-makers as, for example, monitoring procedures and incentive systems, are assumed to mitigate agency problems. Under the assumption that departmental managers seek to maximize their individual utility function managers choose an option according to the expected contribution to utility. Decision-making as well as performance evaluation usually is based on noisy information - which in turn imposes risk on departmental managers (for an overview see [12]).

Whereas design elements related to the organisational structure as well as those directed to the behavioural control of individuals have a long tradition in management research, the interrelations have received less attention. Especially, the extensive findings on contracting under imperfect information are rarely related to organisational design elements [2] - although, the imperfections of information as well as the need for monitoring information and incentive system depend on organisational design elements like differentiation, delegation of decision-making or coordination mode [22, 23].

This paper aims to contribute to bridging the gap and, especially, to analyse interdependencies between organisational design elements related to the organisational structure and those related to individual behavioural control of decision-makers. Given the familiar distinction between decision-facilitating and decision-influencing information as introduced by Demski and Feltham [5], the article concentrates on the former. The research question is the following: *How is the overall performance of organisations affected by imperfections of decision-facilitating information that self-interested departmental decision-makers and the headquarters use under the regime of different organisational design elements and with different complexities of interactions among decisions?* Thus, in order to investigate the research question, a method is required that allows controlling a multitude of issues in interaction with each other. The multitude of interrelated issues would be particularly difficult to control in empirical research and would lead to intractable dimensions in formal modelling. In contrast, a strength of simulation methods is to deal with manifold interdependent issues [4] and with respect to decentralized decision-making as incorporated in the research question an agent-based simulation appears appropriate.

The remainder of this article is organised as follows: subsequently, the architecture of the simulation model is presented. The third part introduces the results of the simulations which are discussed in the fourth part of the article.

2 Model

The simulation model is based on Kauffman's NK model [9, 10] which has been successfully adopted to organisations by management scholars (e.g. [3, 16, 17, 19,

20]). However, the NK model so far has not been extensively employed to analyse decision-making with imperfect information which is in the focus of this paper.

In particular, the research question of this article requires to model two aspects: firstly, an organisational structure with departments, competencies of decision-makers, coordination mechanisms and incentive systems has to be modelled. This part of the model is similar to that introduced by Siggelkow and Rivkin [20]. Secondly, the research question requires to model imperfect information of decision-makers at the departmental and at the headquarters' level. Therefore, this paper applies an advanced version of the NK model in that adaptive walks on noisy fitness landscapes, as introduced by Levitan and Kauffman [13], are modelled. In order to represent imperfect information in organisations our model differs from Levitan and Kauffman [13] in that informational imperfections are differentiated with respect to organisational levels and competencies. Thus, in general, the distinctive feature of the model presented in this article is that a noisy variant of the NK model is applied that differentiates between various forms and allocations of informational imperfections according to the organisational structure.

2.1 Organisational Structure

Our organisations have a ten-dimensional binary decision problem, i.e. they have to make decisions $d_i \in \{0,1\}$ with $i = 1,..10$. Thus, according to the NK model $N = 10$ and each single state of decision d_i provides a contribution C_i with $0 \leq C_i \leq 1$ which follows a uniform distribution to overall performance $V(\mathbf{d})$. Performance contribution C_i does not only depend on the single decision d_i but also on K_i other decisions $d_{j,j\in\{1,...K_i\},j\neq i}$ (thus, unlike the standard NK model [9, 10], the degree of interactions given by K might differ among decisions i). The overall organisational performance $V(\mathbf{d})$ is defined as normalised sum of performance contributions C_i which results in

$$V(\mathbf{d}) = \frac{1}{N}\sum_{i=1}^{N} C_i = \frac{1}{N}\sum_{i=1}^{N} f_i(d_i; d_{j=1}, ..., d_{j=K_i}) \tag{1}$$

where $j \neq i$.

An organisation consists of a headquarters and three departments subscripted by r. Each department has primary control over a subset of the ten decisions (e.g., department a over decisions 1 to 3, department b over decisions 4 to 7 and department c over decisions 8 to 10). Each department head seeks to find the best configuration for the "own" subset of choices assuming that the other departments do not alter their prior subsets of decisions.

Two different modes of coordination are mapped (for these and further modes [20]): either the departments make proposals to the headquarters which chooses that proposal with the highest overall performance or the departments decide autonomously about their "own" partial decision problems. In the latter case the overall configuration of decisions results as a combination of these departmental decisions without any central intervention.

However, what the "best" configuration from a departmental manager's perspective is, clearly depends on the incentives given. In principal, the incentive structure is modelled similar to Siggelkov and Rivkin [20]: each department head is rewarded according to a linear incentive scheme [1] which may consist of two components. Firstly, the compensation depends on the "own" performance P_r^{own}, i.e. those C_i related to that subset of i decisions a department head is in charge of. Due to interactions the "own" performance may also be affected by decisions made in other departments. Secondly, the rewards may also depend on the contributions related to the "residual" decisions, i.e., those decisions other departments are responsible for. This part of the value base of compensation for department r is denoted with $P_r^{residual}$. Of course, in case of cross-departmental interactions a department also is able to affect the residual performance. The relation between own and residual performance in the value base is controlled by a simple parameter: in all simulations the own performance is rewarded whereas the portion of the residual performance is variable and controlled by the parameter INC_r. Thus, if $INC_r = 1$ the department head is given firmwide incentives.

Assuming that departmental managers seek to maximize their compensation, each of them decides in favour of that known alternative subset of decisions which leads to the maximum value base. However, in our model managers dispose over imperfect information.

2.2 Informational Imperfections

In order to integrate informational imperfections noise is imposed on the contributions of decisions to performance differentiated for each decision-maker according to the information quality he/she reasonably disposes of: a common idea in the theories of organisation mentioned in the introduction and others is that decision-makers in organisations dispose of information with different levels of imperfections. For example, departmental decision-makers are assumed to have relatively precise information about their own area of competence, though limited cross-departmental knowledge whereas the headquarters might have rather coarse-grained, but organisation-wide information. Thus, in the model managers decide on the basis of *perceived* performance contributions of the alternatives rather than true performance effects of their decisions. Thus, the *perceived* value base $\tilde{B}_r$ for compensation of a certain decisional configuration that the head of department r assesses is given by

$$\tilde{B}_r(\mathbf{d}) = \left[\tilde{P}_r^{own} + INC_r \times \tilde{P}_r^{residual}\right] / N \tag{2}$$

For reasons of simplicity we assume that the perceived performances result from additive relations between true performances and an error term like

$$\tilde{P}_r^{own} = P_r^{own} + e_r^{own} \tag{3}$$

and

$$\tilde{P}_r^{residual} = P_r^{residual} + e_r^{residual} \tag{4}$$

Obviously, each error term e_r^{own} and $e_r^{residual}$ has to be specified with respect to various dimensions as, for example, its magnitude or functional relation to true performance contributions of decisions. Of course, a functional relation among true performance and noise does not necessarily exist and, even if, countless functional forms are possible (see [13] for an example of decreasing noise with increasing actual performance in the context of molecular adaptation). However, in the area of performance measurement in organisations - at least, due to the accounting systems [11] - it is reasonable that high (low) true values are related with high (low) noise. For simplicity, subsequently, all error terms follow a Gaussian distribution $N(\mu;\sigma)$ with expected value $\mu = 0$ and standard deviation σ as a *relative* error imputed to the true value of performance.

Furthermore, it is reasonable that decision-makers have learning capabilities, i.e., that informational imperfections related to areas in the fitness landscapes already known are smaller than if new configurations of decisions are explored. In order to integrate this aspect a simple counter, denoted with $count_{\mathbf{d}}$, is employed: initialized with 1, whenever a certain configuration of decisions **d** is been realised (either by combination of departmental decisions in coordination mode "decentralized" or by the headquarters' final decision in mode "proposal") $count_{\mathbf{d}}$ is incremented by 1. Thus, when assessing **d** again the error terms are divided by $count_{\mathbf{d}} > 1$. In consequence, the performances as perceived by head of department r are given by

$$\tilde{P}_r^{own} = P_r^{own} \left[1 + \frac{1}{count_{\mathbf{d}}} (N(0;\sigma_r^{own})) \right] \tag{5}$$

and

$$\tilde{P}_r^{residual} = P_r^{residual} \left[1 + \frac{1}{count_{\mathbf{d}}} (N(0;\sigma_r^{residual})) \right] \tag{6}$$

Additionally, the headquarters' informational imperfections have to be specified since in coordination mode "proposal" the headquarters chooses one of the departments' proposals. The headquarters decides on basis of the *perceived* overall organisational performance given by

$$\tilde{V} = V \left[1 + \frac{1}{count_{\mathbf{d}}} (N(0;\sigma^{head})) \right] \tag{7}$$

In the simulations artificial organisations are sent to fitness landscapes in order to search for configurations **d** of decisions with superior levels of performance $V(\mathbf{d})$. Within each time period a department randomly discovers two alternative *partial* configurations of those binary decisions d_i the department is in charge of: an alternative configuration that differs in one decision and a another alternative which differs in two decisions compared to the current configuration. So, together with the status quo each department has three options to choose of and favours that option which is *perceived* to promise the highest value base for rewards.

3 Results

In the simulations, after a "true" fitness landscape is generated, distortions due to the various informational imperfections in the organisation are added. Table 1 shows the scenarios of informational imperfections which are characterized by the error terms according to the previous section.

Table 1 Simulated scenarios of informational imperfections

Scenario	σ_r^{own}	$\sigma_r^{residual}$	σ^{head}
IS^{abc+-}	0.05	0.2	0.12
IS^{abc-+}	0.1	0.15	0.12
IS^{ab+-c-}	dpmts. a,b: 0.05 dpmt. c: 0.2	dpmts. a,b: 0.2 dpmt. c: 0.3	0.12
$\text{IS}^{abc:0}$	0	0	0

Scenario IS^{abc+-} might represent departmental decision-makers specialised in their task whereas in scenario IS^{abc-+} decision-makers are more generalist-like. In scenario IS^{ab+-c-} one department decides on the basis of very noisy information whereas the two other departments are "specialists" as in scenario IS^{abc+-}. Compared to the departments the headquarters disposes of information with medium level of noise which is relevant only in the coordination mode "proposal". Scenario $\text{IS}^{abc:0}$ serves as benchmark in that here all decision-makers have perfect information.

The simulated organisations show one of two "extreme" types of interactions among the decisions each department is in charge of: in a "self-contained" type the intra-departmental interactions among decisions are highly intense whereas no cross-departmental interdependencies occur [7]. This organisational design reflects a situation where loosely coupled divisions seek to enhance their performance dealing with internal interdependencies while the overall performance just is the sum of divisional performances contributions. This case represents a decomposable "block-diagonal" decision problem [17, 3]. In contrast, a structure with "full interdependence" represents the case where all decisions affect the performance contributions of all other decisions. Thus, in this situation the decisional interdependence is raised to the maximum and the departmental decisions show maximum cross-departmental interference. Of course, among these two "extreme" structures countless scenarios with medium cross-departmental interactions are possible.

Tables 2 and 3 report the results of the simulations differentiated by incentive structure and coordination mode. Each row represents 5000 simulations of the respective scenario over 200 periods (for termination with noisy fitness measurements see [13]). Indicator Speed 1 ($V_{t=1} - V_{t=0}$), i.e. performance improvement in the first period, might be regarded as a pure measure for the effects of informational imperfections if compared to perfect information ($IS^{abc:0}$). Speed 2 ($V_{t=2} - V_{t=1}$) gives an indication for the reduction of noise due to knowledge of a certain part of the fitness landscape acquired in the first period. The average performance during the adaptive walk, $\overline{V}_{T200}$, is a condensed measure of the speed of performance enhancements whereas $V_{t=200}$ reflects the performance level that is achieved in period 200.

Table 2 Effects of imperfect information in "self-contained" organisations

Scenario	Speed 1 ($V_{t=1}$ $-V_{t=0}$)	Speed 2 ($V_{t=2}$ $-V_{t=1}$)	Average Perform. ($\overline{V}_{T200}$)	CI[a] of $\overline{V}_{T200}$	Perform. in $t = 200$ ($V_{t=200}$)	CI[a] of $V_{t=200}$
Coordination mode "decentralized"						
Departmental incentives only (INC = 0)						
IS^{abc+-}	0.17882	0.05177	0.98588	±0.00114	0.98840	±0.00115
IS^{abc-+}	0.17075	0.05097	0.98522	±0.00104	0.98872	±0.00105
IS^{ab+-c-}	0.17088	0.04992	0.98412	±0.00106	0.98757	±0.00108
$IS^{abc:0}$	0.18406	0.05146	0.98596	±0.00117	0.98813	±0.00118
Firmwide incentives (INC = 1)						
IS^{abc+-}	0.11219	0.02876	0.90858	±0.00169	0.96314	±0.00293
IS^{abc-+}	0.12813	0.03401	0.94707	±0.00133	0.98316	±0.00151
IS^{ab+-c-}	0.10028	0.02747	0.87241	±0.00183	0.92434	±0.00429
$IS^{abc:0}$	0.18258	0.05122	0.98515	±0.00122	0.98726	±0.00123
Coordination mode "proposal"						
Departmental incentives only (INC = 0)						
IS^{abc+-}	0.09533	0.06124	0.96567	±0.00178	0.97017	±0.00182
IS^{abc-+}	0.09383	0.06004	0.96336	±0.00179	0.96814	±0.00184
IS^{ab+c-}	0.09440	0.06004	0.96387	±0.00177	0.96878	±0.00182
$IS^{abc:0}$	0.12052	0.07246	0.98551	±0.00116	0.98858	±0.00118
Firmwide incentives (INC = 1)						
IS^{abc+-}	0.08288	0.05553	0.95359	±0.00183	0.96255	±0.00193
IS^{abc-+}	0.08609	0.05799	0.95471	±0.00183	0.96260	±0.00192
IS^{ab+-c-}	0.08199	0.05313	0.94969	±0.00189	0.95936	±0.00200
$IS^{abc:0}$	0.11973	0.07345	0.98488	±0.00117	0.98799	±0.00118

[a] Confidence intervals at a confidence level of 0.001

4 Interpretation

Conventional wisdom suggests that decision-making based on imperfect information results in lower performance improvements compared to decision-making with perfect information: obviously, assessing alternatives with imperfect information

Table 3 Effects of imperfect information in "full-interdependent" organisations

Scenario	Speed 1 ($V_{t=1} - V_{t=0}$)	Speed 2 ($V_{t=2} - V_{t=1}$)	Average Perform. ($\overline{V}_{T200}$)	CI[a] of $\overline{V}_{T200}$	Perform. in $t = 200$ ($V_{t=200}$)	CI[a] of $V_{t=200}$
Coordination mode "decentralized"						
Departmental incentives only (INC = 0)						
IS^{abc+-}	0.01851	0.00701	0.72178	±0.00406	0.76519	±0.00746
IS^{abc-+}	0.01530	0.00712	0.70864	±0.00367	0.74507	±0.00739
IS^{ab+-c-}	0.01584	0.00363	0.70830	±0.00365	0.74014	±0.00742
$IS^{abc:0}$	0.02039	0.00476	0.72751	±0.00415	0.76847	±0.00750
Firmwide incentives (INC = 1)						
IS^{abc+-}	0.03787	0.02423	0.87121	±0.00276	0.89703	±0.00277
IS^{abc-+}	0.03973	0.02597	0.87547	±.00265	0.89196	±0.00276
IS^{ab+-c-}	0.03328	0.01938	0.85681	±0.00298	0.89613	±0.00303
$IS^{abc:0}$	0.04768	0.03538	0.87551	±0.00267	0.88092	±0.00273
Coordination mode "proposal"						
Departmental incentives only (INC = 0)						
IS^{abc+-}	0.12013	0.03147	0.85847	±0.00312	0.86175	±0.00321
IS^{abc-+}	0.12110	0.03033	0.85849	±0.00305	0.86184	±0.00314
IS^{ab+-c-}	0.12462	0.03032	0.85948	±0.00308	0.86298	±0.00316
$IS^{abc:0}$	0.14459	0.03078	0.86456	±0.00298	0.86599	±0.00301
Firmwide incentives (INC = 1)						
IS^{abc+-}	0.13185	0.03321	0.86789	±0.00297	0.87093	±0.00303
IS^{abc-+}	0.13042	0.03256	0.86863	±0.00291	0.87139	±0.00296
IS^{ab+-c-}	0.12685	0.03167	0.86526	±0.00298	0.86842	±0.00304
$IS^{abc:0}$	0.15903	0.03206	0.88159	±0.00271	0.88314	±0.00274

[a] Confidence intervals at a confidence level of 0.001

can result in a choice in favour of an alternative which only appears favourable, whereas, in fact, it reduces performance compared to the status quo level. Underestimating the status quo level of performance due to measurement errors might foster the false estimation. In consequence, managers might decide in favour of a "false positive" alternative. Vice versa, in course of the adaptive walk a "false negative" new alternative is rejected because its marginal contribution to performance appears worse than it actually is and, thus, the current configuration of decisions **d** is perpetuated. This situation might be fostered by an overestimation of the status quo level of performance. So, with imperfect information decision-making might go in the wrong direction or, at least, overall performance is enhanced with lower speed compared to decision-making with perfect information.

Tables 2 and 3 provide broad, but *no general* support for conventional wisdom. However, three aspects deserve closer attention and are, subsequently, discussed in greater detail.

Firstly, comparing the three "noisy" scenarios with respect to average performance $\overline{V}_{T200}$ and performance level $V_{t=200}$, neither the "specialist" scenario IS^{abc+-} nor the "generalist-like" scenario IS^{abc-+} generally predominates and, given the confidence intervals, even the scenario IS^{ab+-c-} with one fairly uninformed department is not clearly inferior. Obviously, the ranking with respect to speed of enhancement and achieved level of performance depends on the setting of interaction structure, incentives and coordination.

Secondly, the losses in speed and magnitude of performance enhancements in the course of noisy adaptive walks compared to perfect information subtly vary with incentive structure, coordination mode and interactions among decisions.

The *role of incentives* most clearly appears in the "decentralized" coordination mode - as here the headquarters does not override departmental behaviour. Therefore, to analyse the role of incentives we focus only on the results for the "decentralized" mode reported in the upper parts of tables 2 and 3 respectively and in Fig. 1. In case of a "self-contained" structure with departmental rewards only performance losses due to informational imperfections are negligible (Fig. 1A) - on a high performance level; in contrast, with firm-wide incentives losses in speed and level of performance appear enormous (Fig. 1B). In a full-interdependent structure restricting rewards to departmental performance is disastrous compared to firmwide incentives (Fig. 1C vs. 1D), and, with this misalignment of incentives, informational imperfections appear of minor relevance. Given that decision-makers want to maximize rewards, the incentive system also affects which information is factored in decision-making - in our model whether solely (fairly precise) information about departmental or also (imprecise) information about cross-departmental effects of decisions on performance is used. Thus, the incentive system somehow works as an ambivalent "*gatekeeper*": it controls on the one hand which information and, thereby, which imperfections on the other hand find a way into decision-making. Obviously, in a full-interdependent structure it is important to consider cross-departmental effects. In contrast, in a self-contained structure actually no cross-departmental interactions exist but due to informational imperfections the decision-maker might think they exist. These 'phantom" interactions lead to performance losses which are the more severe the worse the cross-departmental information. These results somewhat correspond to findings of Siggelkow [18] and Bushman [2].

The *role of the coordination mode* can be analysed by comparing the results within Tables 2 and 3 respectively. The analysis suggests that with central coordination the different levels of imperfect information on the departments' sites lead to minor differences in speed and level of performance improvements than in the decentralized mode - although partially at a lower and partially at a higher performance level. Thus, it appears that introducing the headquarters' information-processing power has a *stabilising* function in context of informational imperfections.

Thirdly, in some situations imperfect information leads to *superior overall performance* $V_{t=200}$ than achieved under perfect information. In the reported results this is the case in a full-interdependent structure with decentralized coordination and firmwide incentives (Fig. 1D and Table 3; none of the confidence intervals, CI of $V_{t=200}$ of the "noisy" scenarios overlaps with that one of perfect scenario). This somewhat surprising result is in line with Levitan and Kauffman for adaptive walks

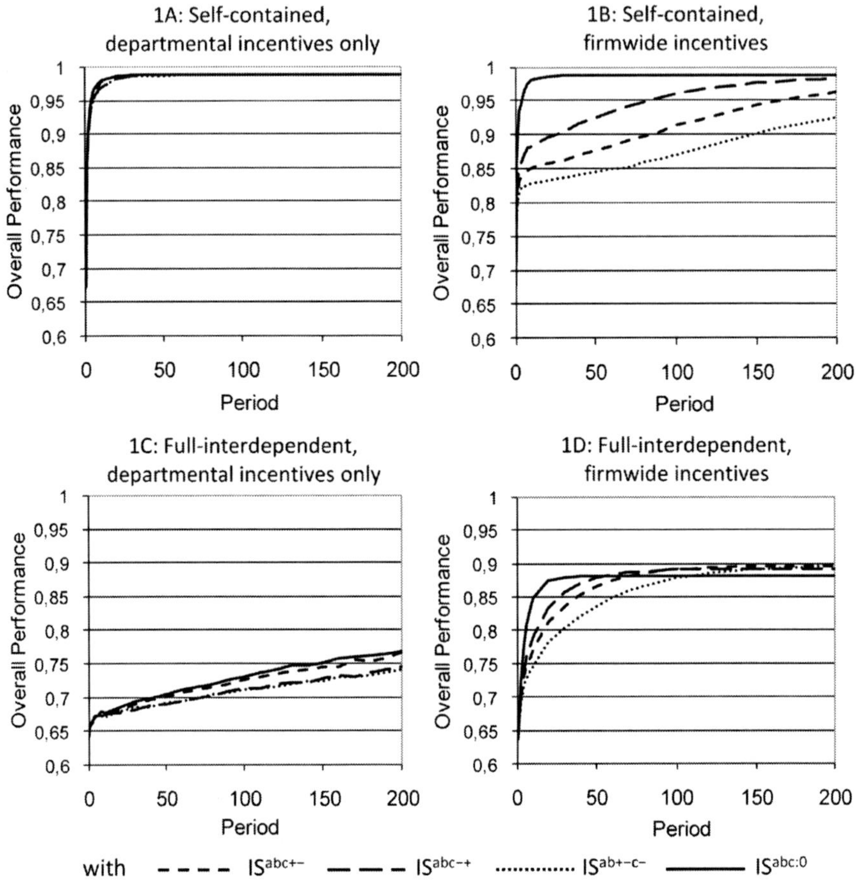

Fig. 1 Adaptive walks in coordination mode "decentralized"

on noisy fitness landscapes: they show that (in the course of molecular adaptations) slight noise can lead to higher performance levels than obtained without measurement noise [13]. With "false positive" decisions an organisation goes to astray which would, of course, not be the case with perfect information; from a short-term "wrong way" there is a chance to discover new superior configurations of decisions. So, informational imperfections may afford the opportunity to leave a local maximum and find a superior configuration of decisions, and, thus, somewhat serve as "*catalysts*" for new solutions. However, the question remains why in the reported results the "beneficial" effects (with non-overlapping confidence intervals) only appear in case of full-interdependence. We argue that with more interactions more local peaks exist and, so, it is more likely that an organisation sticks to a local performance peak (e.g. [17]). In consequence, the chance to leave a local peak via "false positive" assessments is higher when interactions are more intense.

5 Conclusion

The results suggest that the effects of informational imperfections subtly depend on organisational design elements as incentive systems, coordination mode and interactions among the decisions delegated to departments. Moreover, the results might throw some new light on organisational as well as information systems design. Conventionally, imperfect information is assumed to lead to severe performance losses. According to the findings of this paper, in some organisational settings imperfect information appears to cause insignificant losses. So, given that enhancing information quality usually causes additional costs the pay-off of investments in improved information systems might be rather unclear. Moreover, in some situations informational imperfections appear beneficial, presumably by enabling exploration of new solutions. Obviously, this puts ambitions for preferably perfect information in perspective.

References

1. Banker, R.D., Datar, S.M. (1989) Sensitivity, Precision and Linear Aggregation of signals for Performance Evaluation. Journal of Accounting Research **27**(1):21–39
2. Bushman, R.M., Indjejikian, R. J., Smith, A. (1995) Aggregate Performance Measures in Business Unit Manager Compensation: The Role of Intrafirm Interdependencies. Journal of Accounting Research **33**(1):101–129
3. Chang, M.-H., Harrington, J. E. (2006) Agent-Based Models of Organizations. In: Tesfatsion, L., Judd, K. L. (eds.), Handbook of Computational Economics: Agent-Based Computational Economics Vol. 2, 1273–1337. Elsevier, Amsterdam
4. Davis, J.P., Eisenhardt, K.M., Bingham, C.B. (2007) Developing Theory Through Simulation Methods. Academy of Management Science **21**:480–499
5. Demski, J.S., Feltham, G.A. (1976) Cost Determination: A Conceptual Approach. Iowa State University Press, Ames
6. Eisenhardt, K.M. (1989) Agency Theory: An Assessment and Review. Academy of Management Science **14**:57–74
7. Galbraith, J.R. (1973) Designing Complex Organizations. Addison-Wesley, Reading (MA)
8. Ginzberg, M.J. (1980) An Organizational Contingencies View of Accounting and Information Systems Implementation. AOS **5**:369–382
9. Kauffman, S.A. (1993) The Origins of Order: Self-Organization and Selection in Evolution. Oxford University Press, Oxford
10. Kauffman, S.A., Levin, S. (1987) Towards a General Theory of Adaptive Walks on Rugged Landscapes. Journal of Theoretical Biology **128**:11–45
11. Labro, E., Vanhoucke, M. (2007) A Simulation Analysis of Interactions among Errors in Costing Systems. The Accounting Review **82**:939–962
12. Lambert, R.A. (2001) Contracting Theory and Accounting. Journal of Accounting and Economics **32**:3–87
13. Levitan, B. and Kauffman, S. (1995) Adaptive Walks with Noisy Fitness Measurements. Molecular Diversity **1**:53–68
14. March, J.G., Simon, H.A. (1958) Organizations. Wiley, New York
15. Marschak, J., Radner, R. (1972) Economic Theory of Teams. Yale University Press, New Haven and London
16. Rivkin, J.W., Siggelkow, N. (2003) Balancing Search and Stability: Interdependencies Among Elements of Organizational Design. Management Science **49**:290–311

17. Rivkin, J.W., Siggelkow, N. (2007) Patterned Interactions in Complex Systems: Implications for Eploration. Management Science **53**:1068–1085
18. Siggelkow, N. (2002) Misperceiving Interactions Among Complements and Substitutes. Management Science **48**:900–916
19. Siggelkow, N., Levinthal D.A. (2003) Temporarily Divide to Conquer: Centralized, Decentralized, and Reintegrated Organizational Approaches to Exploration and Adaptation. Organization Science **14**:650–669
20. Siggelkow, N., Rivkin, J.W. (2005) Speed and Search: Designing Organizations for Turbulence and Complexity. Organization Science **16**:101–122
21. Simon, H.A.(1976) Administrative Behavior. A Study of Decision-Making Processes in Administrative Organizations, third ed. The Free Press, New York
22. Thompson, J.D. (1967) Organizations in Action. Social Science Bases of Administrative Theory. McGraw-Hill, New York
23. Tushman, M.L. (1979) Work Characteristics and Subunit Communication Structure: A Contingency Analysis. Administrative Science Quarterly **24**:82–98
24. Tushman, M.L., Nadler, D.A. (1978) Information Processing as an Integrating Concept in Organizational Design. Academy of Management Review **3**:613–624

Social Interactions and Innovation: Simulation Based on an Agent-Based Modular Economy

Bin-Tzong Chie and Shu-Heng Chen

Abstract Using an agent-based modular economic model, we study the effect of social interactions on product innovation and its further impact on competitiveness dynamics. Two firms with different intensities of social interactions are placed in a context of duopolistic competition. The macroscopic analysis based on various criteria, including the market share, profit rate, accumulated capital, waste ratio and consumers' satisfaction level, indicates that high social interaction within the firm can lead to not only a healthy firm but also a healthy economy. However, this positive result is undermined by the catastrophic nature of the modular economy as shown in the microscopic analysis. Furthermore, the mesoscopic analysis shows that in the long run the duopoly market tends to become a monopolistic market, and there is a non-trivial probability that the low-interaction firm will drive out the high-interaction firm. The risk of innovation in this model may be greater than what the usual economic model may expect.

1 Introduction: The Purpose of this Study

The purpose of this study is to examine the significance of social interactions to innovation and the competitiveness of firms. While obtaining direct measures of social interactions can be difficult, conceptually speaking, factors ranging from social network topologies (hardware) to language and culture (software) all contribute to social interactions. If the contents of social interactions are *ideas*, then it is not difficult to imagine how these may facilitate innovation processes. In this paper, instead

---

Bin-Tzong Chie
Department of Industrial Economics, Tamkang University, Taiwan
e-mail: chie@mail.tku.edu.tw

Shu-Heng Chen
AI-ECON Research Center, Department of Economics, National Chengchi University, Taiwan
e-mail: chen.shuheng@gmail.com

M. LiCalzi et al. (eds.), *Progress in Artificial Economics*, Lecture Notes in Economics and Mathematical Systems 645, DOI 10.1007/978-3-642-13947-5_11,

of a direct modeling of the mechanism generating social interactions, we take this as exogenously given and adopt a conventional approach frequently used in the early agent-based economic literature to parameterize social interactions [2, 4, 11]. Concretely speaking, we use the parameter crossover rate from evolutionary computation (more specifically, genetic programming) as a proxy for the degree of social interaction. The crossover rate has been considered to be the most important parameter in genetic programming [14], and its significance has been studied in various applications, including finance [9]. However, its possible role in agent-based economic models has only been explored [1, 3] in a limited manner; even in the case of individual learning [18], this parameter is normally set as being homogeneous among all agents. This paper, therefore, takes a closer look at the possible implications of the heterogeneity of this parameter in the context of duopolistic competition.

However, our study is not based on a standard agent-based duopolistic (oligopolistic) model [8, 10, 16], where the product is homogeneous among firms. In those conventional settings, social interactions as idea generators are limited to only pricing strategies or marketing strategies. While marketing and pricing are crucial, the fundamental advantages of competition come from products, and their *varieties* and *novelties*. These elements are, however, missing in earlier agent-based oligopolistic models, but have recently gained attention among agent-based models of innovation [12]. One of these recent developments is to build the innovation model upon Herbert Simon view's of a complex system [17]. To do so, the idea of *modularity* is incorporated into the innovation process and the design of the product can be endowed with a hierarchical modular structure. This is called the *agent-based modular economy* [6, 7].

The agent-based modular economy serves our current purpose well. It has non-trivial innovation (non-price innovation) as its core. The advantages of firms can be gained by not just smart marketing but, more importantly, by *new products*. It is, in this context, that we study the significance of social interactions.

The rest of the paper is organized as follows. Section 2 reviews the foundations upon which the modular economy is built. We present a sketch of how Herbert Simon's notion of modularity can be used to provide a novel representation of the fundamentals of an economy, namely, preferences and products. Section 3 gives a brief description of the simulation design. Section 4 presents the simulation results, and we also put forward macroscopic, microscopic and mesoscopic views of the competitiveness dynamics, followed by the concluding remarks in Sect. 5.

2 Foundation of the Work

The foundation of the work is very much inspired by Herbert Simon in his work on modularity [5] and uses *context-free language*, from computation theory, to represent consumers' preferences and products as *hierarchical trees*. The modular structures of preferences and products are naturally introduced to the fundamentals of the economy through these representations. This modeling enables us to more clearly

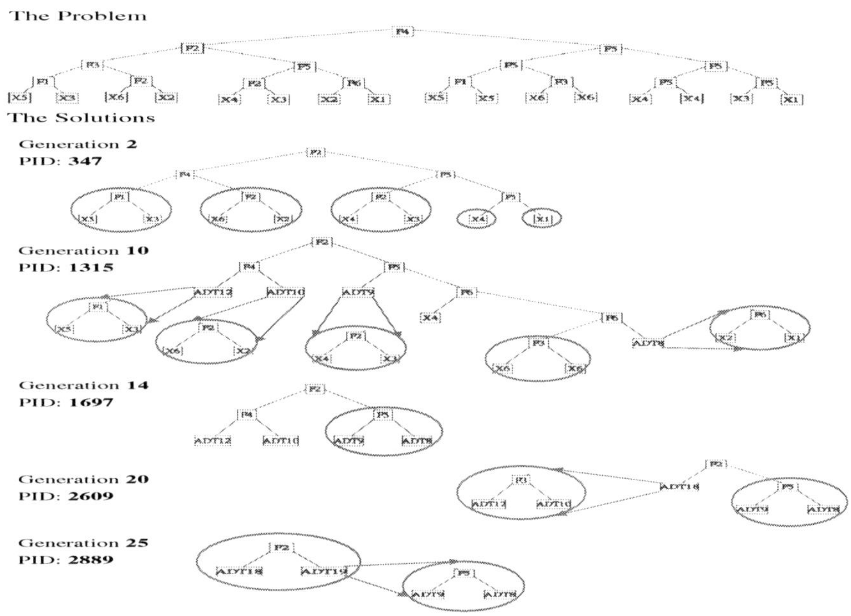

Fig. 1 An illustration of a process of product innovation

see the combinatoric nature of economic problems. *Basically, firms are driven to search for better products in terms of matching consumers' preferences with higher and higher hierarchies.* The whole process itself shows how the economy evolves from simple to complex, and from a quantity-oriented economy to a quality-oriented economy.

What is also presented in this economy is Alfred Marshall's central idea of evolution, i.e., *differentiation* and *integration*. Given the heterogeneity of consumers' preferences, firms are led to be heterogeneous, too. They differ in terms of the varieties of products supplied, and also differ in the degree of specialization or diversification.[1] Since genetic programming can work on the context-free grammar directly, it is used to model the search and evolution process of firms and the economy. A sketch of the innovation process is given in Fig. 1.

On the top of this figure is the parse-tree representation of a consumer's preference, which introduces the "problem" for producers. The whole parse tree corresponds to a *word* generated by recursively applying a number of production rules from a context-free grammar. Each subtree of it is also a word but with less deep recursion. The shallowest recursion is just the application of a single production rule: null $\rightarrow x$, where $x \in \Sigma$, and Σ is the terminal set. All the nodes at the bottom of the tree, also called leaves, are this kind of shallowest recursion. Then the deeper the recursion, the bigger the subtree, and the complete tree involves the largest number of

[1] For some illustrations, see [6], pp. 174-176.

recursions compared to all its subtrees. This recursive structure denotes the modular structure of a preference, from a very primitive preference with shallow recursion to a complex preference with deep recursion. It also indicates the growing and evolving nature of the preference.

Notice that the consumer himself may not know his own preference *ex ante*. In this case, the preference is revealed only up to the extent that there is a product which matches it. This has been shown in the sequence of diagrams below the consumer's preference (tree). In each of these diagrams, one can see that a number of products (all the red circled ones) have been successfully developed to match some subtrees of the whole preference. In this case, only these parts of the preference (the matched ones) are revealed.

3 Simulation Design

Table 1 summarizes the values of the key parameters of this agent-based modular economy.[2] The simulation design in this paper is almost the same as [7] except for the following. Both firms use the modular structure in their innovation and production; technically speaking, the automatically defined terminals [15] are employed by both firms. However, the two firms differ in their organization culture, which results in different degrees of social interaction. In this paper, to make this difference sharp, the high-interaction firm is associated with a crossover rate of 90%, and the low-interaction firm is associated with a crossover rate of 45%.

Generations 100 runs of the simulation are conducted, and each run lasts for 5,000 iterations (generations). Each generation is composed of a number of market days (in our case, five). After each learning cycle, the firm has to decide what to produce, including some new products developed via innovation, how many to produce, and how much to charge. The decision regarding production and R&D is based on the sales and profits statistics collected on the previous market days (the previous generation). The firm then supplies what has been produced plus some new items during the next few market days (the next generation).

Market Day One single market day is described as follows. On each trading day, each consumer will enter the market in an order which is determined *randomly*. In addition, a search intensity, which is exogenously determined, is assigned to each consumer. The search intensity determines the fraction of sellers which the consumer can reach. For example, if search intensity is 50%, then the consumer can reach half of the sellers in the market. If there are two sellers, then the consumer can visit one of the two. Without losing generality, consider now that the ith consumer enters the market. He visits a number of firms determined by the search intensity.

[2] A full understanding of this table requires a lengthy description of the agent-based modular economy, which includes the behavior of firms (cost structure, working capital, and strategies on product diversity, pricing, and R&D), the behavior of consumers (searching intensity, procurement), and the matching process between consumers and producers. Due to space limitations, the interested reader is referred to [6, 7].

Table 1 Parameter Settings

Parameter	Type (Variable)	Range	Default Value
	Producer		
Number of Firms	Integer (n_p)	$[1, \infty)$	2
Initial Working Capital	Integer (K_0)	$[1, \infty)$	500
Working Capital per Gen.	Integer (K)	$[1, \infty)$	500
Inventory Adjustment Rate	Real (λ)	$[0, 1]$	80%
Mark-up Rate	Real (η)	$[0, \infty)$	100%
R&D Rate	Real ($\gamma_{\text{R\&D}}$)	$[0, 1]$	1%
R&D Ceiling	Real ($\overline{\text{R\&D}}$)	$[0, \infty)$	500
Cost per Node	Real (c)	$[0, \infty)$	1.0
	Consumer		
Number of Consumers	Integer (n_c)	$[1, \infty)$	100
Consumer Income per Gen.	Integer (I)	$[1, \infty)$	10000
Depth of Consumer Preference	Integer (d_p)	$[1, \infty]$	6
Depth of Common Preference	Integer (d_c)	$[1, d_p]$	5
Base of Preference to Utility	Integer (z)	$[2, 10]$	4
Price to Utility Ratio	Real (v)	$[0, \infty)$	5.0
Search Intensity	Real (r_s)	$[0, 1]$	100%
	Genetic Operator		
Number of Primitives	Integer (ρ)	$[1, \infty)$	5
Initial Tree Depth	Integer (d_{ini})	$[1, \infty]$	5
Maximum Tree Depth	Integer ($d_{\max}$)	$[1, \infty]$	11
Tournament Size Ratio	Real (r_{ts})	$[0, 1]$	10%
Crossover Rate	Real (p_c)	$[0, 1]$	(1) 45% (2) 90%
Mutation Rate	Real (p_m)	$[0, 1]$	80%
Automatically Defined Terminal	Boolean (ADT)	T, F	T
	Time Schedule		
Market Days per Gen.	Integer (Day)	$[1, \infty)$	5
Number of Generations (Gen.)	Integer (Gen)	$[1, \infty)$	5000

Note: (1) and (2) are the parameters for the low-social interaction firm and high-social interaction firm, respectively.

For all firms which he visits, he examines all commodities available, and calculates the consumer's surplus for each commodity. Based on that, he ranks the commodities from the one with the highest consumer's surplus to the lowest, and then picks the one with the highest surplus subject to his affordability. He then leaves the market, and the $i+1$th consumer enters the market, and repeats the same procedure. This process continues until all consumers have entered the market, and then this market day is over. The calendar then turns to the next market day.

Our focus in the simulation is to see the firms' competitiveness dynamics. The initial attempt here is to measure competitiveness by market share. However, since the two firms differ in terms of their products supplied, a *comprehensive* market share covering all products sold will be used. Let $sales_t^L$ ($sales_t^H$) be the total sales of all products produced by the low-interaction (high-interaction) firm in the tth period (the tth generation). Let mkt_t^L (mkt_t^H) be the corresponding market share of the low (high)-interaction firm: $mkt_t^L = sales_t^L/(sales_t^L + sales_t^H)$,

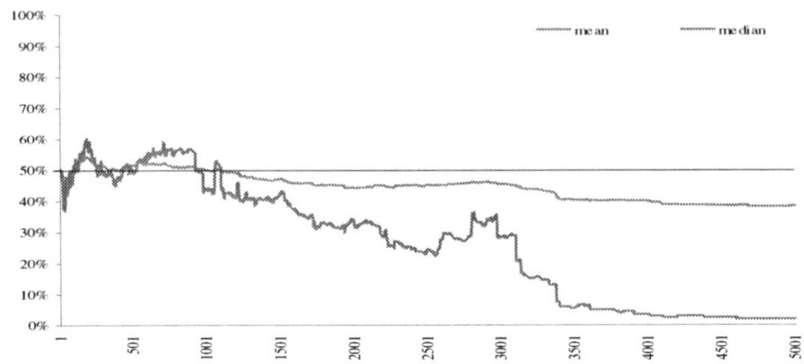

Fig. 2 Social Interactions and Competitiveness of Firms

$mkt_t^H = sales_t^H/(sales_t^L + sales_t^H)$; obviously, $mkt_{i,t}^L + mkt_{i,t}^H = 1$. The results of a specific run are further indexed by i: $mkt_{i,t}^L$ and $mkt_{i,t}^H$.

4 Simulation Results

4.1 Competitiveness Dynamics: A Macroscopic View

Our presentation of the simulation results starts with the market-share dynamics or the competitiveness dynamics. Let $\overline{mkt_t}$ and $\widetilde{mkt_t}$ be the mean and median of $mkt_{i,t}^L$ over the 100 runs.

$$\overline{mkt_t} = \frac{\sum_{i=1}^{100} mkt_{i,t}^L}{100}, \quad \widetilde{mkt_t} = median\{mkt_{i,t}^L\}_{i=1}^{100}, \quad t = 1,\ldots,5000. \tag{1}$$

These two statistics are then drawn in Fig. 2. The line in the middle is the 50% line, where the two firms are tied. Any deviation of the $x\%$ from this 50% line indicates the superiority of one firm over the other.

Both curves have clearly shown that the firm with high interaction will eventually take the lead. What is particularly appealing is that, if we focus on the medium ($\widetilde{mkt_t}$), it clearly shows that the firm with low interaction will become extinct in the long run. Hence, the duopolistic competition is so keen that the pace of innovation is vital for the firms' survivability. The firm with the high interacting environment, which may facilitate innovation, will eventually drive out the rival firm (the low-interaction firm). This result seems to be well anticipated.

Nonetheless, $\overline{mkt_t}$ behaves quite differently. While it also has a tendency to decline, the speed is rather slow, and in the end it almost stagnates at the level of 40%. Therefore, the sharp difference between $\overline{mkt_t}$ and $\widetilde{mkt_t}$ indicates something intricate and less well anticipated. Basically, this sharp difference implies that the high-interaction firm did not always win. In fact, based on our 100 runs,

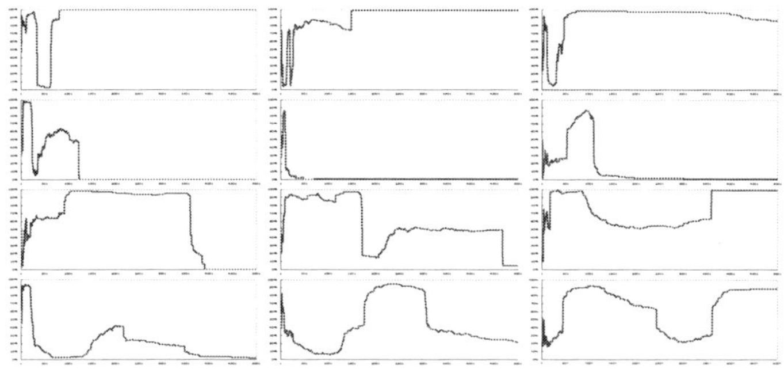

Fig. 3 Competitiveness (Market-Share) Dynamics: Some Single Runs

$mkt^H_{i,5000} > mkt^L_{i,5000}$ in 61 out of the 100 runs. In other words, there exists a non-trivial probability (39%) that the high-interaction firm will be dominated. This result can be regarded as a manifestation of the risk of innovation in general and the risk of more frequent innovation in this specific context. To obtain a feel for this risk, it is, therefore, necessary to take a finer look at some individual runs.

4.2 Competitiveness Dynamics: A Microscopic View

The multiple runs of the duopolistic simulation enable us to inquire into the possibility of various emerging patterns which may interest us when we want to get a closer look at the risk of innovation. To do this, Fig. 3 presents a glossary of the $mkt^H_{i,t}$ of 12 individual runs. While this is only one eighth of our total number of runs, it is sufficient to highlight a few points to be made below. First, we would like to know whether the high-interaction firm can *easily* or *quickly* dominate the market and then give its rival firm no chance to return; in the extreme, $mkt^H_{i,t}$ converges to one quickly. We call this pattern *an easy win*. Cells (1,1) (Run 57) and (1,2) (Run 60) of Fig. 3 demonstrate the series of $mkt^H_{i,t}$ which is closer to this pattern: the higher interaction firm dominated the market in the early stage and became the monopoly all the way to the end. However, even for these two cases, there is great uncertainty in the early stage of the market, as shown by the severe fluctuation of $mkt^H_{i,t}$. In a few periods, $mkt^H_{i,t}$ even fell almost to the floor (0). So, even for these two runs, the battle is not that easy. Cell (1,3) (Run 19) exhibits a similar pattern, but it never completely drove out the rival firm and in the later stage it started to lose many of its customers. Cells (2,1) (Run 28), (2,2) (Run 29) and (2,3) (Run 63) reveal the opposite of the easy-win pattern: after initial fluctuations, the high-interaction firm was driven out.

The six cases above are evidence that the dominating firm is determined by the fierce competition in the early stage. It is a pattern of a harsh kick-off followed by an easy win or lose. It is very hard to get stabilized in the early stage; a market-share reversal can happen quite easily at this stage, caused by only a one-step movement.

The transition to the full dominance ($mkt^H = (\approx)1$) or the extinction ($mkt^H = (\approx)0$) is not smooth at all. The behavior of these six cases naturally motivates us to ask the next question. Would it then be possible that a well-established firm would be presented with a sudden detrimental shock which causes its quick death? We call this pattern *a full reversal*. Cell (3,1) (Run 14) provides evidence of a positive answer for a full reversal. In this case, the high-interaction firm already took a market share of 90% for a duration of 3,000 periods, but then it lost all its markets within the next 500 periods, and was no longer able to come back.[3] Hence, this run shows that even a well-established firm is not free from risk. Not only does our modular economy present firms with a tough kick-off, but it also gives no guarantee that the well-established firm will stay.

If a well-established firm still faces the risk of becoming extinct quickly, then we need not mention a long-held balanced competitive situation. This is exactly what we see from Cells (3,2) (Run 12) and (3,3) (Run 3). In Cell (3,2), the balanced situation ($mkt^H \approx mkt^H \approx 1/2$) had remained for more than 2,000 periods until a sudden event caused mkt^H almost drop down to the floor. In Cell (3,3), we see the opposite. The two firms had been in a close range for 2,500 periods before the high-interaction firm suddenly drove out the low interaction firm ($mkt^H = 1, mkt^L = 0$).

A common property shared by the nine cells above is a phenomenon familiarly known as *catastrophes*: a sudden movement made by firms results in their quick triumph or extinction. Simply put, $mkt^{H,L}$ can move to one or zero in a rather short span of time regardless of its current position. While catastrophes can happen quite easily in this modular economy, they do not rule out the possibility of a smooth or slow transition. In Cell (4,1) (Run 59), the high-interaction firm was constantly in an inferior position, but it took 3,000 periods to see its gradual disappearance. Nevertheless, generally speaking, catastrophes, rather than smooth transitions, are the rule.

The ten individual simulations shown so far all tend to indicate that either $mkt_t^H \rightarrow 1$ or $\mathcal{N}(1)$ or $mkt_t^H \rightarrow 0$ or $\mathcal{N}(0)$, where $\mathcal{N}(1)$ and $\mathcal{N}(0)$ are small neighborhoods of 1 and 0, respectively. There are, however, cases showing the wandering behavior of mkt_t^H. Cells (4,2) (Run 100) and (4,3) (Run 7) are cases in point. In both of these two cells, mkt_t^H wandered over a wide range, from 10% or 20% to 90%.

4.3 Competitiveness Dynamics: A Mesoscopic View

Given the catastrophic nature of this modular economy, if $mkt_t^{L,H}$ can wildly switch between zero and one, then it certainly has an important implication for the concentration of the industry; for example, most of the time, the industry can essentially be

[3] Of course, we cannot exclude the possibility that it would have come back had we run the simulation for longer.

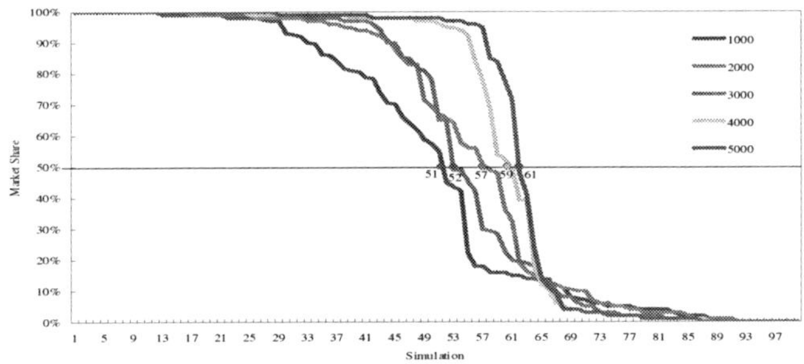

Fig. 4 Ranked Market Share of the High-Interaction Firms

The five plots above are the non-increasing series $\{mkt^H_{j,t}\}_{j=1}^{100}$ obtained by rearranging the original $\{mkt^H_{i,t}\}_{i=1}^{100}$ in descending order: $mkt^H_{1,t} \geq mkt^H_{2,t} \geq ...mkt^H_{j,t} \geq mkt^H_{j+1,t} \geq ... \geq mkt^H_{100,t}$. Five periods of time are sampled: $t = 1000, 2000, 3000, 4000$, and 5000.

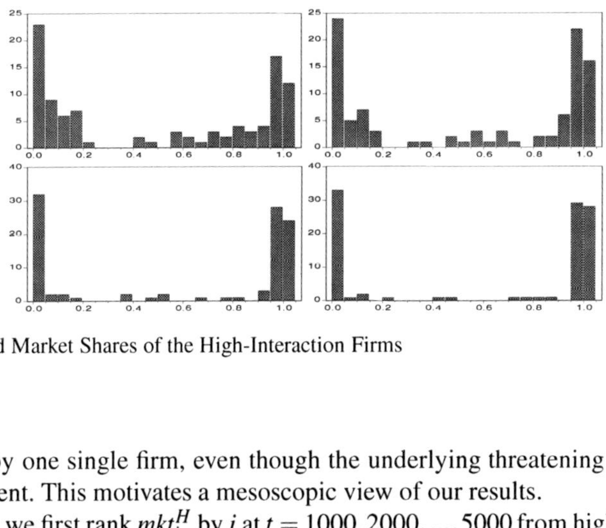

Fig. 5 Ranked Market Shares of the High-Interaction Firms

controlled by one single firm, even though the underlying threatening competition is never absent. This motivates a mesoscopic view of our results.

To do so, we first rank $mkt^H_{i,t}$ by i at $t = 1000, 2000, ..., 5000$ from high to low, and transform the original series $\{mkt^H_{i,t}\}_{i=1}^{100}$ into a non-increasing series $\{mkt^H_{j,t}\}_{j=1}^{100}$, where $mkt^H_{1,t} \geq mkt^H_{2,t} \geq ...mkt^H_{j,t} \geq mkt^H_{j+1,t} \geq ... \geq mkt^H_{100,t}$. We then draw these transformed series in Fig. 4, and they appear in the form of an inverted S curve. The interesting thing is that the curve starts with a long flat section corresponding to the ceiling $mkt^H_t = 1$, and ends with another long flat section corresponding to $mkt^H_t = 0$. The connection between two flat sections becomes steeper and steeper as time moves on, indicating a bimodal distribution with modes in 1 (or $\mathcal{N}(1)$) and 0 (or $\mathcal{N}(0)$). Hence, in very little time, the market is a true duopoly.

Fig. 5 is the histogram of the series $mkt^H_{i,t}$ for four sampling periods t=1000, 2000, 4000 and 5000. We see that the distribution of mkt^H_t tends to converge to

Table 2 Mean Performances of Firms for Other Production Criteria

Social Interaction	ROR	AC	WR
Low	23.08%	63,632,334	40.05%
($p_c = 45\%$)	(57.98%)	(56,388,693)	(30.15%)
High	44.52%	83,984,012	28.76%
($p_c = 90\%$)	(54.08%)	(56,054,398)	(28.11%)
p-value	0.0037	0.0056	0.9966

The three criteria "ROR", "AC", and "WR" refer to the rate of return, accumulated capital and the waste ratio, respectively. The second and the fourth rows are the means of these three criteria over 100 runs. The third and the fifth rows (values inside the parentheses) are the respective standard deviations. The last row gives the p-value of the null hypothesis that the mean of the low-interaction firm is higher than the mean of the high-interaction firm.

a zero-or-one distribution (Kolmogorov's zero-one law or tail event). Hence, the winner does tend to take all, but the winner is not necessarily the high-interaction firm.[4]

4.4 Other Performance Criteria

In addition to market share, it would be useful to look at other possible performances to determine whether they are consistent with market share. What interests us most is the efficiency or the social surplus of the economy. This can be measured by both the producers' surplus and the consumers' surplus. The producers' surplus can be measured by a profit criterion. Here, we use the *rate of return* (ROR), i.e., the profit divided by production capital.[5] In addition to ROR, we also watch the accumulated capital (AC) up to the end of the simulation. Finally, in this economy, not all products can fit the consumers' needs at an acceptable price or they are simply over supplied. In this case, these unsold products completely lose their value in the next period; therefore, many products are wasted in this sense. We also measure the waste ratio (WR) by dividing the wasted unites units produced by the total units produced.

Table 2 summarizes the performance of these criteria. Since we have 100 independent runs, we tabulate the means taken over all these 100 runs. From Table 2, we can see that the high-interaction firm in general performs better than the low-interaction firm. The former has a higher rate of return and a higher accumulated capital, whereas the latter has a higher waste ratio. The differences in these three criteria are all statistically significant. Hence, all three criteria lead to the consistent result that we obtained earlier using the market-share criterion.

4 Of course, to see whether this is the long-term outcome of our modular economy, we need to run the simulation for much longer than we have done now. We shall leave this to a further study.

5 At each period of time, the firm will use only part of its working capital for production purposes (including R&D and advertising). To ensure that the operation is sustained, the firm will also use part of the working capital, called the reserved capital, as a buffer for various market risks. Therefore, production capital is obtained by subtracting reserve capital from working capital.

Table 3 Regression of Consumers' Satisfaction on Market Share

	Mean_Utility ($\bar{U}_{i,t}$)	Max_Utility ($U^{max}_{i,t}$)
mkt^H	12.5065	10.7379
(p-value)	(0.000)	(0.000)
Constant	46.8349	77.2403
(p-value)	(0.000)	(0.000)
R^2	0.049	0.022

Finally, we also examine the performance from the consumers' side. In this economy, we have 100 consumers with heterogeneous preferences. How well have they been served by the duopolistic firm? One simple way to answer this question would be to examine how consumers' satisfaction (realized utility) is affected by the market share of the high-interaction firm. Let $U_{l,i,t}$ be the satisfaction level of the lth consumer it the tth period in the ith simulation. Here, we run a regression of the consumers' satisfaction on the market share of the high-interaction firm ($mkt^H_{i,t}$). We consider two dependent variables: the mean satisfaction ($\bar{U}_{i,t}$) and the maximum satisfaction ($U^{max}_{i,t}$), where

$$\bar{U}_{i,t} = \frac{\sum_{l=1}^{100} U_{l,i,t}}{100}, \quad U^{max}_{i,t} = \max\{U_{l,i,t}\}_{l=1}^{100}. \tag{2}$$

We run the regression by pooling all 100 simulations together which provides data with 500,000 observations (100 runs each lasting for 5,000 generations), and the results are shown in Table 3. The result shows that consumers' satisfaction can be enhanced by the market share of the high-interaction firm, although the latter has only limited explanatory power.

5 Concluding Remarks

Using genetic programming, we simulate an artificial modular duopolistic economy with different social interactions (organizational cultures) which may have an impact on firms' innovation capacity and competitiveness. While at a macroscopic level a higher innovation capacity caused by more intense social interaction can enhance firms' competitiveness, the risk associated with innovation itself is so significant that one can hardly be sure of what to expect on any single occasion. As shown in our microscopic analysis, this idiosyncratic uncertainty is so dramatic that we cannot even recommend investing in firms with high social interaction and hence high innovation, for example, like Google. The essential lesson from our simulation is that the potential risk of this creative culture may be underestimated. What is even worse is that developing an early-warning system is also infeasible given the catastrophic nature of this economy.

Acknowledgements The version has been substantially expanded and revised in light of two anonymous referees' very helpful reviews, for which we are most grateful. The NSC grant 98-2410-H-004-045-MY3 is also gratefully acknowledged.

References

1. Arifovic, J. (1994) Genetic Algorithm Learning and the Cobweb Model. Journal of Economic Dynamics and Control **18** (1): 3–28
2. Birchenhall, C.R. (1995) Modular Technical Change and Genetic Algorithms. Computational Economics **8**: 233–253
3. Birchenhall, C.R., Lin, J.S. (2002) Learning and Convergence to Pareto Optimality. In: Chen, S.H. (ed.), Genetic Algorithms and Genetic Programming in Computational Finance. Kluwer Academic Publishers, Boston
4. Brenner, T. (1998) Can Evolutionary Algorithms Describe Learning Processes? Journal of Evolutionary Economics, **8**: 271–283
5. Callebaut, W., Rasskin-Gutman, D.(eds.) (2005) Modularity: Understanding the Development and Evolution of Natural Complex Systems. The MIT Press, Cambridge (MA)
6. Chen, S.-H, Chie, B.-T. (2005) A Functional Modularity Approach to Agent-Based Modeling of the Evolution of Technology. In: Namatame, A., Kaizouji, T., Aruka, Y. (eds.), The Complex Networks of Economic Interactions: Essays in Agent-Based Economics and Econophysics, LNMES 567, Springer, 165–178
7. Chen, S.-H, Chie, B.-T. (2007) Modularity, product innovation, and consumer satisfaction: An agent-based approach. In: Yin, H., Tino, P., Corchado, E., Byrne, W., Yao, X. (eds.), Intelligent Data Engineering and Automated Learning, LNCS 4881, Springer, 1053–1062
8. Chen, S.-H., Ni, C.-C. (2000) Simulating the Ecology of Oligopoly Games with Genetic Algorithms. Knowledge and Information Systems: An International Journal **2**: 310–339
9. Chen, S.-H., Kuo, T.-W. (2003) Overfitting or Poor Learning: A Critique of Current Financial Applications of GP. In: Ryan, C., Soule, T., Keijzer, M., Tsang, E., Poli, R., Costa, E. (eds.), Genetic Programming, LNCS 2610, Springer, 34–46
10. Curzon Price, T. (1997) Using Co-Evolutionary Programming to Simulate Strategic Behaviour in Markets. Journal of Evolutionary Economics **7**(3): 219–254
11. Dawid, H. (1999) Adaptive Learning by Genetic Algorithms: Analytical Results and Applications to Economic Models, Springer
12. Dawid, H. (2006) Agent-Based Models of Innovation and Technological Change. In: Tesfatsion, L., Judd, K.L. (eds.), Handbook of Computational Economics vol. 2. Elsevier, Amsterdam, 438–475
13. Goldberg, D.E. (2002) The Design of Innovation: Lessons from and for Competent Genetic Algorithms. Kluwer Academic Publishers, Boston
14. Koza, J.R. (1992), Genetic Programming: On the Programming of Computers by the Means of Natural Selection. The MIT Press, Cambridge (MA)
15. Koza, J.R. (1994) Genetic Programming II–Automatic Discovery of Reusable Programs, The MIT Press, Cambridge (MA)
16. Marks, R.E., Midgley, D.F., Cooper, L.G. (2006) Co-Evolving Better Strategies in Oligopolistic Price Wars. In Rennard, J.-P. (ed.), Handbook of Research on Nature-Inspired Computing for Economy and Management, Ch. 52. Idea Group Inc., Hershey (PA)
17. Simon, H. (1965) The Architecture of Complexity. General Systems **10**: 63–76
18. Vriend, N.J. (2000) An Illustration of the Essential Difference between Individual and Social Learning, and its Consequences for Computational Analyses. Journal of Economic Dynamics and Control **24**: 1–19

Threshold Rule and Scaling Behavior in a Multi-Agent Supply Chain

Valerio Lacagnina and Davide Provenzano

Abstract In this paper an agent-based model of self organized criticality is developed in a network economy characterized by lead time and a threshold behavior of firms. Instead of considering the aggregate production of the economy as a whole, we focus on both the propagation and amplification effects of a demand shock in the sectorial productions of a multi-agent supply chain. We study a static network structure representing a relation of firms in a lower-upper stream in an industrial organization. In our model, the individual (R,nQ) policies play an important role in generating a propagation effect across the different layers of the economy, and the propagation turns into the large fluctuations and amplifications of sectorial productions.

1 Introduction

The phenomenon of amplified variations of demand moving upstream in the supply chain (the further away from the consumer) is well known in industrial economics as the bullwhip effect. This effect leads to inefficiencies in supply chains since it increases the cost for logistics, in terms of greater safety stocks and inventory, and lowers the competitive ability of the firms in the network. It is also responsible for low economic performances of the firms because an organization has to cope with the ramifications of failed fulfillments which can lead to contract penalties. Lead time of information and materials and batch ordering are among the main causes for the bullwhip effect. We explicitly introduce these two features in our model of a

Valerio Lacagnina
Dipartimento di Scienze Statistiche e Matematiche "Silvio Vianelli", University of Palermo, Italy
e-mail: `ricopa@unipa.it`

Davide Provenzano
Dipartimento di Scienze Statistiche e Matematiche "Silvio Vianelli", University of Palermo, Italy
e-mail: `provenzano@unipa.it`

M. LiCalzi et al. (eds.), *Progress in Artificial Economics*, Lecture Notes in Economics and Mathematical Systems 645, DOI 10.1007/978-3-642-13947-5_12,

multi-agent supply chain where M firms are organized in K production levels. Each level is composed of the same number L of firms located in different geographical areas. Exogenous independent shocks in the final good demand cause some of the firms to adjust. Their adjustments induce further adjustments of other firms, which in turn induce even further adjustments, and so on. This chain reaction constitutes a propagation mechanism. Each firm buys inputs from a subset of closest firms in the upper level (suppliers) and sells its output to a subset of closest firms in the lower level (customers). The bottom level consists of retailers, meaning firms that sell in the consumer market, whereas the top level consists of firms that have direct access to an unlimited source of primary production inputs. Firms in the network implement a (R, nQ) echelon stock policy operating as follows: if during the economic activity the inventory position goes below or equals the reorder point R, the firm orders the input and starts the production of nQ units of output. Q is the firm's production batch and n is the minimum integer required to increase the inventory position to above the reorder point R. Altogether, from the order placed to the suppliers to the delivery of the output, it takes τ time units for the firm to fulfill customers' demand. Therefore, productive units face stochasticity, nonlinearity and lead time, where the first refers to the independent demand fluctuations for the final good, the second to non-convexities in the firmsâ response to the sectorial shocks and the third to the order processing time. The combination of these three special features results in significant fluctuations in aggregate production and in amplifications of the orders of input moving upstream in the supply chain.

Our study is strongly inspired by Bak et al. (1993), Battiston et al. (2007), Delli Gatti et al. (2005), and Weisbuch and Battiston (2007) in determining the role of local random events on the distribution of production and order dynamics.
In Bak et al. (1993) production networks are defined by edges that represent supplier/customer connections among firms engaged in a production/distribution activity. The authors describe the production avalanches triggered by random independent demand events at the output boundary of the production network.
Battiston et al. (2007) present a simple model of a production network where edges represent inter-firm relationships involving extension of trade-credit. In their model, the domino effect occurs as a consequence of a chain of failures in the ability of firms to reimburse debt.
The paper by Delli Gatti et al. (2005), instead, is a multi-agent model of financial fragility that incorporates the indirect interaction among firms taking place through the endogenous determination of the interest rate on bank loans. When bankruptcies occur, non-refunded loans negatively affect banks net worth and, consequently, credit supply. The reduction in credit supply impacts on the lending interest rate all other firms have to pay hampering their solvency.
Finally, Weisbuch and Battiston study the consequences of simple local processes of orders/production (with or without local failure)/delivery/profit/investment on the global dynamics: evolution of global production and wealth in connection to their distribution and spatial patterns.

Our paper, as well as the above mentioned ones, belongs to the practice of combining heterogeneity and interactions to explain non-normal distributions, scaling

behavior or the occurrence of large aggregate fluctuations generated by small idiosyncratic shocks. In this approach based on *heterogeneous interacting agents* (HIA), a coherent framework is provided by the agent-based methodology we use to carry out our study. In the spirit of complex systems analysis, our aim is not to present specific economic predictions but, primarily, to reproduce the stylized facts common to a large class of models of production networks with a special attention to the effect that lead time and batch ordering generate in the dynamics of the sectorial productions. In fact, although threshold rules are widely observed in individual economic behavior, few theoretical works have shown the relevance of threshold behaviors in aggregation when the shocks are independent across individual units.
The rest of the paper is organized as follows. Section 2 introduces the model. Section 3 outlines the simulation results with particular attention to the distribution of the aggregate. Section 4 concludes the paper.

2 The Model

2.1 The Network Structure

We consider a network economy consisting of M producers/warehouses and K production levels, Fig. 1. The bottom level, $i = 1$, represents firms selling only one type of final good in the consumer market (retailers) whereas the top level, $i = K$, represents firms having direct access to primary production inputs. For each production level output is qualitatively different from input. The total number of firms is constant over time and each stage is composed of the same number $L = M/K$ of firms. Moreover, to avoid border effects, firms are arranged on a cylinder with three input and three output connections of equal importance each. Therefore, the structure of the economy can be represented by a cylindrical lattice where each unit $m \in M$ has coordinates (i, j) (with $i = 1, 2, \ldots, K$ and $j = 1, 2, \ldots, L$).
Time runs discretely in periods $t = 1, 2, \ldots, T$. At the beginning of each period, the retailers' level is buffeted by exogenous random shocks which can take the value zero with probability 1-p and the value one with probability p, where p is a small positive number[1]. One after another, firms at each stage i and at each position j in the stage react to orders coming from downstream decreasing their limited inventory. Therefore, for a generic firm m, the quantity $Y^s_m(t)$ of goods supplied at time t is a minimum between its inventory capacity $I_m(t)$ and orders $\Omega_{C_m}(t)$ coming from its (finite) set C_m of nearest clients one layer below

$$Y^s_m(t) = \min(I_m(t), \Omega_{C_m}(t)) \tag{1}$$

[1] p varies with the size of the economy and, therefore, it should be written as $p(L)$. The mean number of final good orders per period $N(t)$ is then equal to $p(L)L$. For much information about the value of p, see [1].

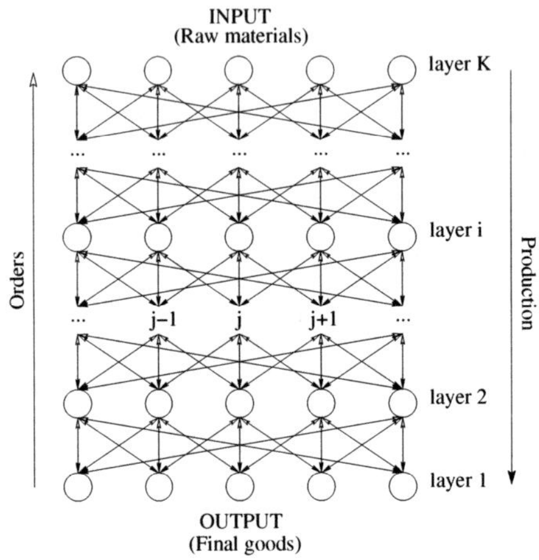

Fig. 1 The network of firms

where $\Omega_{C_m}(t) = \sum_{c \in C_m} Y_c^d(t)$.

If a producer m experiences a shortage of goods or its inventory reduces under a specific threshold R_m, it determines the desired production $\Phi_m(t)$, the units of input $Y_m^d(t)$ needed to start production and, if $i < K$, it transfers evenly distributed orders to its (finite) set S_m of nearest suppliers one layer above[2]. The units of input demanded to supplier $s \in S_m$ will then be equal to:

$$Y_m^{d,s}(t) = \frac{Y_m^d(t)}{|S_m|}. \tag{2}$$

Layer $i = K$ is instead replenished by an unlimited source of primary input outside the boundary of the lattice. We assume the relation between the desired output and the units of input to be expressed by the function $Y_m^d(t) = q\Phi_m(t)$, where q is further taken equal to 1 without loss of generality[3].

Once the upstream flow of orders stops, production starts and proceeds down the network one layer after the other towards the bottom level where the final good is sold to consumers. Retailers ($i = 1$) fill out orders specified as exogenous shocks outside the system. Each of these is assumed to be either zero or one.

Technology of stage $i, \forall i \in K$, locally minimizes its average cost at the output levels 0 and nQ, with n being an integer and Q the minimum production batch. In particular:

$$\Phi_m(t) = \begin{cases} 0 & \text{if } \Omega_{C_m}(t) < I_m(t) - R_m \\ nQ & \text{otherwise} \end{cases} \tag{3}$$

[2] In the proposed model $|C_m| = |S_m| = 3$.

[3] We assume that once input is received, the production process immediately starts.

We let the generic firm m use capital $A_m(t)$ as the only input to its production technology, and the maximum number of batches be a percentage θ of its capital: $n \leq \theta A_m(t)$. Capital productivity θ is constant and uniform across firms, and the capital stock never depreciates. We assume a positive deterministic processing time τ_i for level i and we denote the units of goods produced at time t in the same level with:

$$P_i(t) = \sum_{j=1}^{L} \Phi_{i,j}(t - \tau_i). \tag{4}$$

Two are the consequences of the processing time: orders to be filled out by production are backlogged and a distribution policy is needed for a firm to decide how to distribute its insufficient inventory among the received orders. Backlog at time t is set equal to

$$B_m(t) = \Omega_{C_m}(t) - Y^s_m(t) \tag{5}$$

while, when orders can not be fully satisfied, a firm distributes its inventory to the customers proportionally with their orders. Therefore, the quantity of goods supplied by firm m to its customer c at time t will be equal to

$$Y^{s,c}_m(t) = I_m(t) O_c(t) \tag{6}$$

with $O_c(t) = \dfrac{Y^d_c(t)/|S_c|}{\Omega_{C_m}(t)}$.

In this way, for any firm m, $Y^{s,c}_m(t) \leq Y^{d,m}_c(t)$ is always satisfied with equality holding when the inventory is sufficient to fill out the orders coming from downstream. The inventory law of motion looks as follows:

$$I_m(t) = I_m(t-1) + \Phi_m(t - \tau_i) - Y^s_m(t). \tag{7}$$

2.2 Capital Flows

Agents in the lattice are also connected one another through a flow of payments. In fact, each delivery of goods results into payments from the customer to the supplier. To keep things simple, we abstract from issues of pricing and we suppose firms in each level i to be placed in L different islands so as to consider them as different markets[4]. Since arbitrage opportunities across islands are imperfect, the individual selling price for firm m's output is the random outcome of a market process around the average industry price of output $p_i(t)$, according to the law $p_m(t) = p_i(t)u_j(t)$ where the relative price $u_j(t)$ for the output of the single firm j, is a random variable uniformly distributed in $[1-\xi, 1+\xi]$ and independent of $p_i(t)$. For the average industry price the following relation holds:

$$p_i(t) = p(t) + (K - i + 1) \tag{8}$$

where $p(t) = 1$ is the price of the primary production input.
Each supplier has to pay, in addition to the input cost, a cost $c(t)$ per produced item, a cost r per item in the reorder level R[5], a penalty λ per unit of unfilled demand and a significant positive cost ρ for carrying inventory in excess of the reorder point R plus a production batch Q. We simplify the model and the notation by supposing that the manufacturing cost $c(t)$ is constant over time. For each firm, profits are the difference between the valued quantity of delivered products and the just mentioned costs

$$\Pi_m(t) = p_m(t)Y_m^s(t) - \sum_{s \in S_m} p_s(t)Y_m^{d,s}(t) - c\Phi_m(t) - rR_m - \lambda B_m(t) - \rho ExI_m(t) \tag{9}$$

where $ExI_m(t) = I_m(t) - (R_m + Q_m)$. We suppose that all profits are reinvested into the firms. Production capacity of all firms is thus upgraded (or downgraded in case of negative profits) according to the law

$$A_m(t+1) = A_m(t) + \Pi_m(t). \tag{10}$$

2.3 From Bankruptcy to Rebirth

We suppose that firms whose capital is not sufficient to cover fixed costs go into bankruptcy, their production capacity goes to zero, and they stop both production and delivery. Moreover, when a firm leaves the market, it deletes both the orders made to its suppliers and the orders received from its customers. Suppliers, on their turn, are not allowed to stop the production made to cover the demand of the just bankrupted firm experiencing, eventually, an extra inventory for the produced but not delivered goods. Customers of the bankrupted firm, instead, undergo a shortage of goods if the quantity of intermediates provided by the lasting suppliers is not sufficient to start the production process. The extra-inventory cost and the backlog cost decrease firms' profit and hamper their financial stability. After a latency period, a re-birth process occurs in the corresponding vertex for a new firm which exploits the

[4] The hypothesis of firms producing the same good on different markets is actually equivalent to the hypothesis of monopolistic firms producing differentiated goods on the same market.

[5] The cost $r \times R$ can be thought as a fixed cost per unit of time to continue an existing firm.

business opportunity to produce for the downstream neighborhood of the previously bankrupted firm. New firms are created at a fraction of the average firms capital.

3 Simulation Results

3.1 Parameters Choice

Unless otherwise stated, the following results are obtained for a production network with $L = 2000$ nodes per layer and $K = 5$ layers between the input and the output. Simulations are run for $T = 5000$ time steps that, assuming an average number of 250 working days per year, correspond to 20 years of activity. A single time step is composed of two phases: an upward flow of orders and a downward flow of production across the network, plus updating capital according to profits.

Initial capital and initial inventory are uniformly and randomly distributed among firms with

$$A_m(0) = A_{init} \in [10,11] \tag{11}$$

and

$$I_m(0) = I_{init} \in [1,4]. \tag{12}$$

For all the firms in the network R and Q, meaning the reorder point and the production batch, are constant over time and equal to 1 and 3 units of goods respectively. Therefore, the maximum quantity of goods a firm can stock in its inventory without incurring in an extra inventory cost is equal to 4 units. The extra inventory cost ρ is 0.2 per unit of good in excess, the penalty λ is 0.05 per unit of unfilled demand, and the cost per unit of good in the reorder point (fixed) is 0.15. Variable production cost c is 0.3 per produced unit and the price interval width ξ is 0.2. Finally, capital productivity θ equals 1 for each firm of the network.
In case of bankruptcy of a firm the inactivity time is set as $\delta = 1$ time step and the rebirth capital is assumed to be proportional to the average firm capital: $A_{entry} = 10^{-1}\overline{A}$.

3.2 Firm Size Distribution

Histogram in Fig. 2 shows the distribution of firms' size, as measured by their capital, at the end of the time horizon when the processing time τ is equal to 1 time step. It is plain that the final distribution of firm is characterized by a lot of small firms and very few performing ones.

Since a first (but not unique!) signal of power law behavior is the presence of a linear relationship in the upper tail of a distribution on the log-log scale (the so-called Zipf plot), we then plot the cumulative distributions of firms' capital resulting

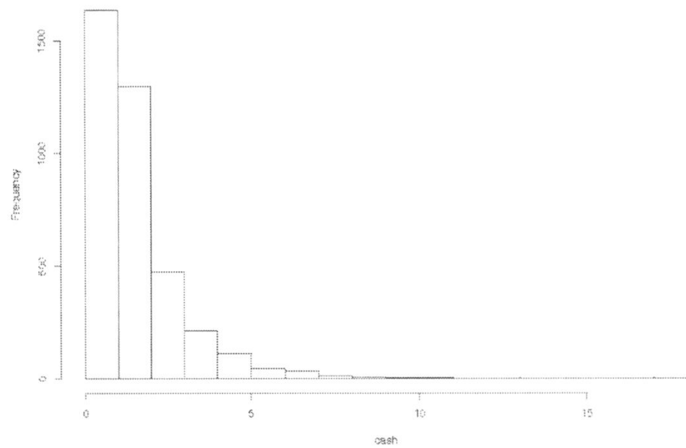

Fig. 2 Firm size distribution at the end of the time horizon ($T = 5000$) when the processing time equals 1 time unit

from our simulations. Figure 3 definitely shows a clear linearity in the right tail of the simulated firm size distribution at the end of the time horizon .

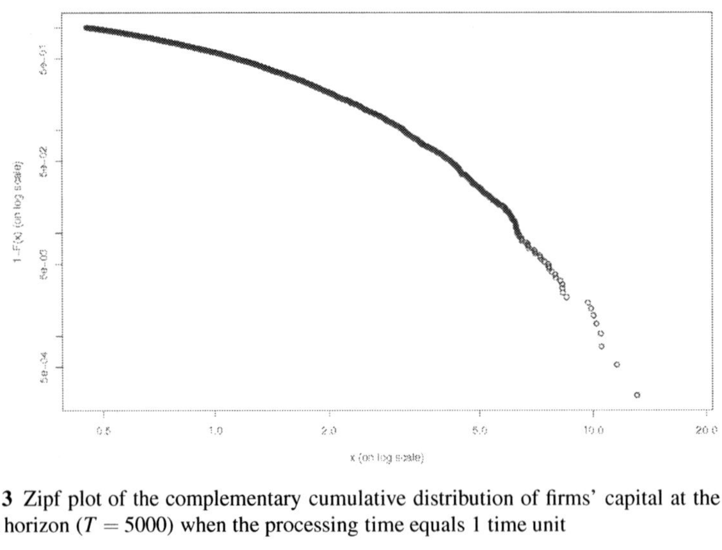

Fig. 3 Zipf plot of the complementary cumulative distribution of firms' capital at the end of the time horizon ($T = 5000$) when the processing time equals 1 time unit

Finally, to confirm the power law hypothesis, we also look for the scalability of sums, which is a mathematical/geometrical property of random variables with power law tails to show a power law behavior even after aggregation. From a graphical point of view it is sufficient to aggregate the original dataset, net of the empirical mean, several times by factors of two and plot each new aggregated sample on the Zipf graph.[6]. Qualitatively, the presence of a scaling behavior can be noticed looking at the right tails of the plotted distributions that should be approximately parallel with similar slopes. Figure 4 shows all this for our data.

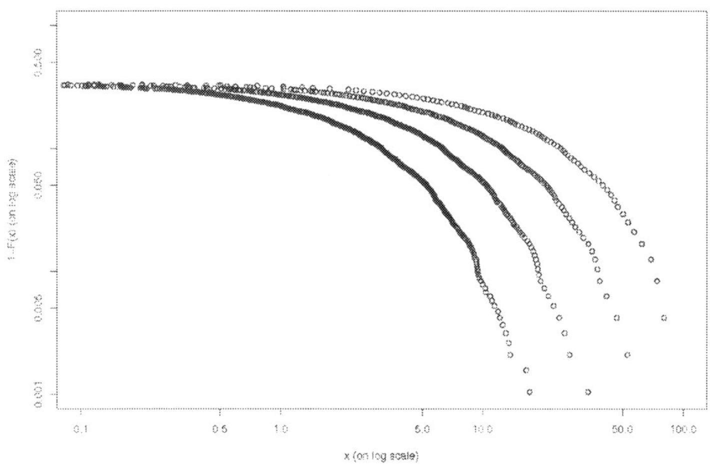

Fig. 4 Graphical representation of the scaling property at the end of the time horizon ($T = 5000$) when the processing time equals 1 time unit

Figure 5 shows the orders variability in the supply chain for the last 250 days of simulation when the processing time equals 1 time unit. Moving from the first (bottom) to the last (top) layer the range of variation in the orders fluctuation increases revealing the presence of the phenomenon known as the bullwhip effect. When the production process takes 3 time units (not shown in the present paper) orders variability range is even higher. We then plot, Fig. 6, the time evolution of the bankruptcies in two different simulations: processing time equal to 1 (top) and 3 (bottom) time units. In both graphs the first 100 periods have been deleted to get rid of transients. It is plain that the higher processing time increases the costs for the agents and, therefore, their financial fragility.

Indeed, in the bottom plot bankruptcies are anticipate, wider, and more frequent.

---

[6] For a clear description of the aggregation process see [3].

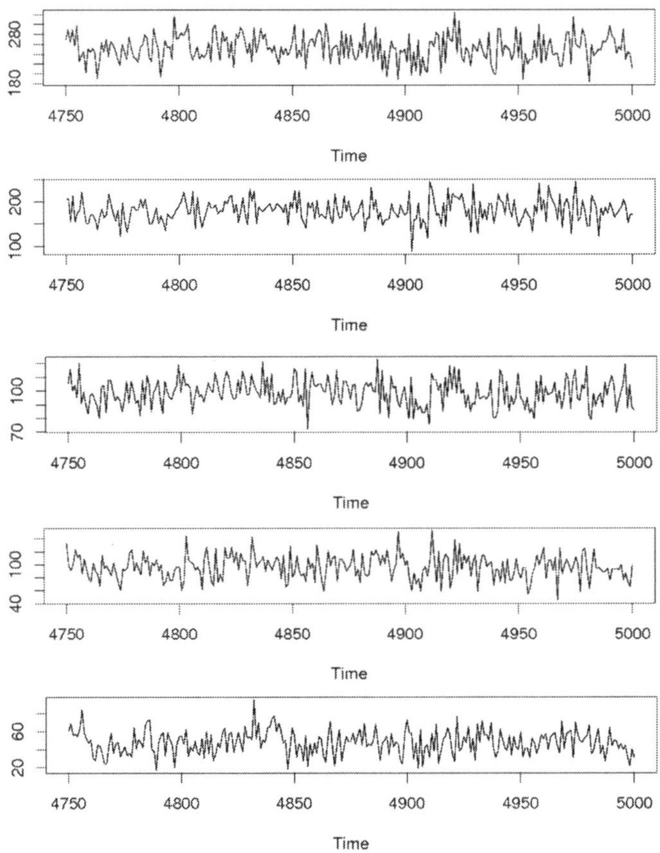

Fig. 5 Time series of orders moving from the first (*bottom*) to the last (*top*) production layer

4 Conclusions

In this paper we analyze a (R, nQ) economy with finite agents characterized by lead time and a threshold adjustment policy. In our framework firms are arranged in a cylindrical lattice and are connected one another by a supplier/customer activity. Firms in the lattice have a structure of costs that takes into account the economic effects that lead time and batch ordering can generate on their financial solidity. In fact, lead time and the batch ordering generate sectorial dynamics which amplify orders moving upwards in the supply chain. These amplifications introduce significant positive costs for the firms when they generate an extra inventory or unfilled orders. Our framework can therefore yield avalanches of bankruptcies when the fluctuations hamper the financial solidity of the firms. In the spirit of complex systems analysis, our aim is not to present specific economic predictions but, primarily, to reproduce the stylized facts common to a large class of models of production networks. In

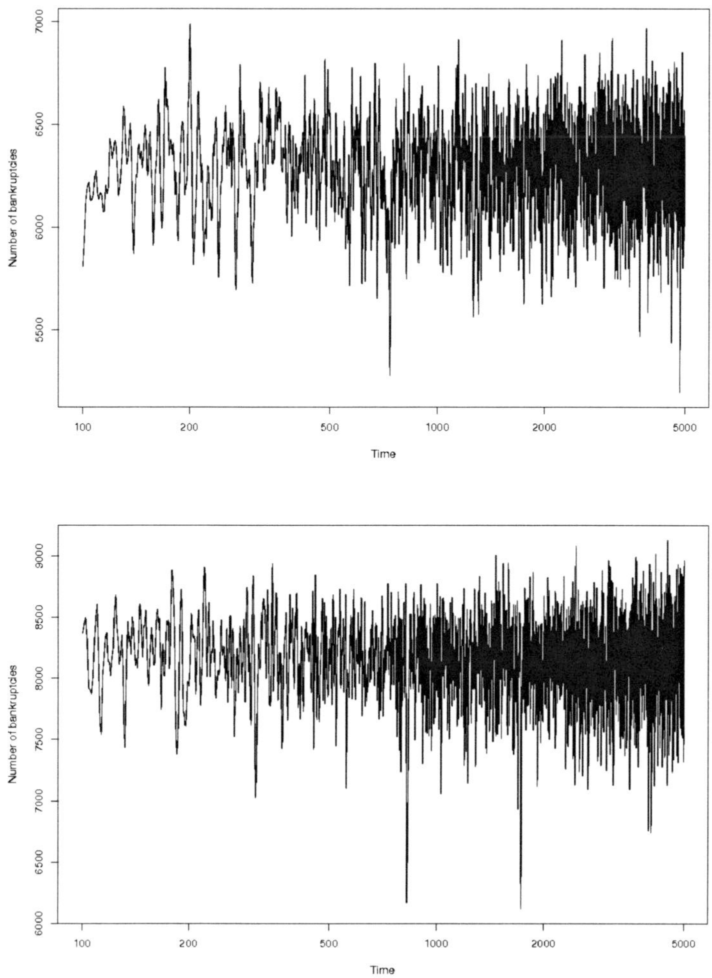

Fig. 6 Time evolution of the number of bankruptcies in the production system when the processing time is 1 (*first plot*) or 3 (*second plot*) time units

fact, our model reproduces qualitatively the main stylized facts of industrial demography, such as firm size distribution and the spatio-temporal patterns for output and bankruptcies. In particular, the distribution of the firms' size is shown for an economy where production takes 1 time step whereas the effect of sectorial dynamics are shown also in the case of a production process of 3 days.

References

1. Bak P, Chen K, Scheinkman J, Woodford M (1993), Aggregate fluctuations from independent sectoral shocks: self-organized criticality in a model of production and inventory dynamics. Ricerche Economiche 47:3–30
2. Battiston S, Delli Gatti D, Gallegati M, Greenwald B, Stiglitz J E (2007), Credit chains and bankruptcy propagation in production networks. Journal of Economic Dynamics and Control 31(6):2061–2084
3. Cirillo P, Jürg Hüsler (2009), On the upper tail of Italian firms' size distribution. Physica A: Statistical Mechanics and its Applications 335:1546–1554
4. Clauset A, Shalizi CR, Newman MEJ (2009), Power-law distributions in empirical data. arXiv: 0706.1062v1 [physics.data-an]
5. Clementi F, Gallegati M (2005), Pareto's Law of Income Distribution: Evidence for Germany, the United Kingdom, and the United States. In: Chatterjee A, Yarlagadda S, Chakrabarti BK (ed) Econophysics of Wealth Distributions. Springer, Milan, doi: 10.1007/88-470-0389-X
6. Delli Gatti D, Di Guilmi C, Gaffeo E, Giulioni G, Gallegati M, Palestrini A (2005), A new approach to business fluctuations: heterogeneous interacting agents, scaling laws and financial fragility. Journal of Economic Behavior and Organization 56(4):489–512
7. Drăgulescu A, Yakovenko VM (2001), Evidence for the exponential distribution of income in the USA. The European Physical Journal B 20(4):585–589
8. Engwall L (1968), Size distributions of firms-a stochastic model. Swedish Journal of Economics 70:138–159
9. Patelli P (2006), Nonlinear Dynamical Systems in Economics. In: Del Santo PP (ed) Universality of Nonclassical Nonlinearity. Springer, New York
10. Stanley MHR, Buldyreva SV, Havlinb S, Mantegna RN, Salingerc MA, Stanley H E (1995), Zipf plots and the size distribution of firms. Economics Letters 49(4):453–457
11. Weisbuch G, Battiston S (2007), From production networks to geographical economics. Journal of Economic Behavior and Organization 64:448–469
12. Fujiwara Y (2004), Zipf law in firms bankruptcy. Physica A: Statistical and Theoretical Physics 337(1-2):219–230

Part V
Industry Sectors

Information and Search on the Housing Market: An Agent-Based Model

John Mc Breen, Florence Goffette-Nagot, and Pablo Jensen

Abstract We simulate a closed rental housing market with search and matching frictions, in which both landlord and tenant agents are imperfectly informed of the characteristics of the market. Landlords decide what rent to post based on the expected effect of the rent on the time-on-the-market (TOM) required to find a tenant. Each tenant observes his idiosyncratic preference for a random offer and decides whether to accept the offer or continue searching, based on their imperfect knowledge on the distribution of offered rents. The steady state to which the simulation evolves shows price dispersion, nonzero search times and vacancies. We further assess the effects of altering the level of information for landlords. Landlords are better off when they have less information. In that case they underestimate the TOM and so the steady-state of the market moves to higher rents. However, when landlords with different levels of information are present on the market, the better informed are consistently better off. The model setup allows the analysis of market dynamics. It is observed that dynamic shocks to the discount rate can provoke overshoots in rent adjustments due in part to landlords use of outdated information in their rent posting decision.

John Mc Breen
ENS-LYON, Laboratoire de Physique, UMR 5672, Lyon, F-69007, France and
LET, UMR 5593, University of Lyon, France e-mail: john.mcbreen@let.ish-lyon.cnrs.fr

Florence Goffette-Nagot
Université de Lyon, Université Lyon 2, F - 69007, Lyon, France. CNRS, GATE Lyon-St Etienne, UMR 5824, 69130 Ecully, France. e-mail: goffette-nagot@gate.cnrs.fr

Pablo Jensen
ENS-LYON, Laboratoire de Physique, UMR 5672, Lyon, F-69007, France and
LET, UMR 5593, University of Lyon, France e-mail: p.jensen@ens-lyon.fr

M. LiCalzi et al. (eds.), *Progress in Artificial Economics*, Lecture Notes in Economics and Mathematical Systems 645, DOI 10.1007/978-3-642-13947-5_13,

1 Introduction

In the urban rental housing market two categories of agents meet. The first category consists of landlords who post rents. The second category consists of tenants who choose among offers. These markets are imperfect as can be seen by the existence of vacancies, price dispersion and nonzero search times for all agents. One source of imperfection is that both categories of agents are imperfectly informed about the characteristics of the market, and acquiring information is costly. These imperfections (in comparison to a theoretical perfectly competitive market) are referred to as search frictions. They has been extensively studied in search-matching models of the labor market, see [12] for a review. [16] created a model of the owner-occupier housing market in which buyers are also sellers and the cost of search effort and its efficiency are defined by an exogenous matching function. This model was later extended to a spatial rental market by [6]. The 'thin' nature of the housing rental market due to the heterogeneity of housing and tenants' idiosyncratic tastes has been used to explain vacancies by [2].

The contribution of the present paper is to propose a multi-agent model as a basis for relaxing many of the assumptions of analytical models in order to obtain a more realistic dynamic model of housing markets. We develop a simulation model of a closed urban housing market focusing on the role of information. In particular, we examine changes in the steady state configuration due to alterations in the level of information available to agents. The major results concern the different influences of the level of information for landlords. Landlords are penalised when they are better informed, as when they are less well-informed underestimations of the TOM move the market to higher rents which are accepted by the tenants. However, when landlords are heterogeneous in information, the better informed are better off.

It has been shown in [13] that price dynamics appear to be led by changes in the vacancy rate. More precisely natural vacancy rates are crucial in determining the strong correlations between fluctuations in the vacancy rate and the evolution of rents. Numerous authors report similar results including [8] also in the rental housing market, [14, 9] in the office rental market and [10] in the purchasing housing market. In a rental market for single family homes, higher asking rents have been shown to lead to longer TOMs [1].

The role of information in the evolution of housing markets has received much attention. [7] studied the correlations in prices and liquidity changes over the housing market cycle. [5] examined a number of possible explanations of the correlations in price and liquidity changes, and found evidence supporting sellers slow rate for updating their beliefs.

[4, 3] use agent-based simulations to relax the assumption of a single price with random matching and Nash bargaining in two rental housing models. These static models examine the distributional effects of rent controls [3] and 'access discrimination' [4], modelled as a reduced matching probability. These analyses however do not explicitly model agents' search behaviour.

2 Model

In our rental housing search model, landlords post rents that become take-it-or-leave-it offers to the tenants. Landlords face a trade-off between setting a higher rent and finding a tenant more quickly. Their optimising behaviour is based on their knowledge of the market state, in terms of TOM for different asking rents.

Tenants are supposed to observe a sample of the offer distribution and to visit one randomly chosen residence each iteration. They accept offers based on an optimising behaviour that trade-offs a quicker match, and therefore reduced search costs, against a lower rent. Tenants must decide their reservation utility U_{res}, the minimum utility they are willing to accept from a residence. They obtain a utility flow U_h from the market, after experiencing a negative utility flow U_s^T while searching. Tenant agents' have a separable utility function whose housing part differs depending on their situation on the housing market. When housed, the agent's instantaneous utility flow from 'housing' is given by: $U_h^T = Y - R + \eta \geq 0$ where U_h^T is the instantaneous utility flow, Y is housing budget, R is the rent paid and η is the idiosyncratic preference of an agent for an apartment that is discovered by the agent once the apartment is visited. $\eta \sim N(0, \sigma)$, where σ is the variance of the normally distributed idiosyncratic preferences. σ is a percentage of the housing budget, see Table1. While searching, the agent's instantaneous utility flow from 'housing' is $U_s^T = Y - C_T < 0$ where C_T is the monetarised cost of searching.

Each tenant has a *housing budget* Y that is drawn from a uniform distribution in the range [100,198]. A searching agent sees one randomly chosen apartment from the distribution of offers and must decide whether to accept it or keep searching. This will depend upon their idiosyncratic preference for the apartment, the posted rent, their housing budget, the cost of search and their outside opportunity. That is the value of continuing to search, which they assess based on their knowledge of a percentage S_T of the full distribution of offers.

U_{res} is optimised to yield the maximum utility per unit time over the expected period of search and residence. Tenant agents idiosyncratic preferences do not play a role in this decision. The expected benefit per unit time for a given reservation utility is

$$B_T\left(U_{res}\right) = \frac{U_s^T}{X + T\left(U_{res}\right)} T\left(U_{res}\right) + \frac{\mathrm{E}\left[U_h^T(U_{res})\right]}{X + T\left(U_{res}\right)} X \tag{1}$$

where U_s^T is the utility flow experienced while searching, X is the expected duration of residence, $T\left(U_{res}\right)$ is the expected search time and $\mathrm{E}\left[U_h^T\left(U_{res}\right)\right]$ is the expected utility flow per iteration once housed. Agents assume the market state to be constant.

The expected search time $T\left(U_{res}\right)$ is equal to the number of residences seen divided by the number of acceptable residences. $\mathrm{E}\left[U_h^T\left(U_{res}\right)\right]$ is given by the average utility of residences, which are expected to yield utilities larger than U_{res}. Once a residence is rejected it cannot be revisited, unless it remains vacant and is randomly reselected. All tenant agents participate in the housing market and searching tenant agents recalculate their reservation utility each iteration.

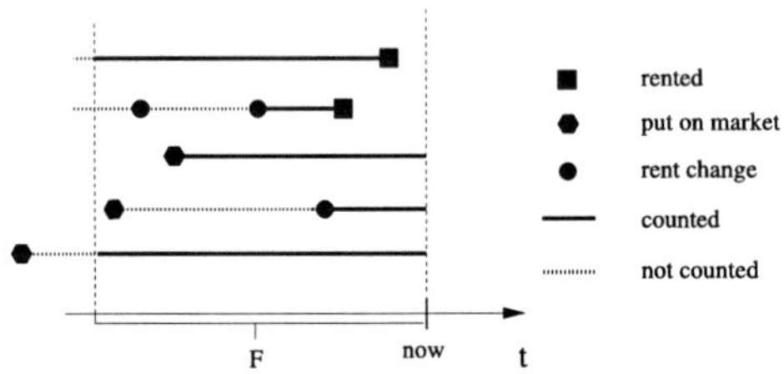

Fig. 1 The periods of time-on-the-market known to landlords

Landlords have three possible states, having a tenant, being on the market and being off the market. The corresponding utilities are, $U^L_{occ} = R - C_L$ the rent minus the maintenance cost, $U^L_{vac} = -C_L$, and 0 their outside opportunity. Landlords' decide what rent will give, in expectation, the highest benefit per iteration,

$$B_L(R) = \frac{-C_L}{X + T(R)} T(R) + \frac{R - C_L}{X + T(R)} X \tag{2}$$

where C_L is the maintenance cost and $T(R)$ is the expected time-on-the-market.

Landlords must estimate the relationship between TOM and expected rent flows. They are assumed to have access to information on a certain percentage of residences on the market over the last F iterations. They know for these residences how many iterations they were on the market at their most recent market price, within the last F iterations. They also know whether or not they have been rented, as shown in Fig. 1. The above procedure generates two histograms, one of the cumulative times on the market within each rent interval, and one of the number of sales within each interval. This allows landlords to calculate the probability per iteration of finding a tenant for a range of rent intervals, implicitly assuming that the probability to sell was constant. This probability is simply the number of agreed rents divided by the cumulative times on the market. The landlords estimate the best least-squares fit of the exponential function for the expected TOM $T(R)$ as a function of the rent R.[1] [2] $T(R) = \alpha \exp(\beta R)$, where α and β are fit parameters. Figure 2 shows an example fit of $T(R)$ and the corresponding expected profits. Landlords review their decisions with probability $1/F$.

Simulation Procedure

- Searchers visit a randomly chosen apartment, and accept or reject it.
- A portion of landlords $(1/F)$ whose apartments remain vacant decide if they shall change their rent or withdraw from the market.
- A portion of landlords $(1/F)$ who have withdrawn from the market decide if they shall return.

[1] Each point $T(R) = \omega$ is given a weight equal to the natural logarithm of the number of rentals $N(R)$ in the rent interval centered on R, plus one, that is $weight = \ln(N(R) + 1)$.

[2] Asking rents cannot be more than 10% above the highest agreed rent seen.

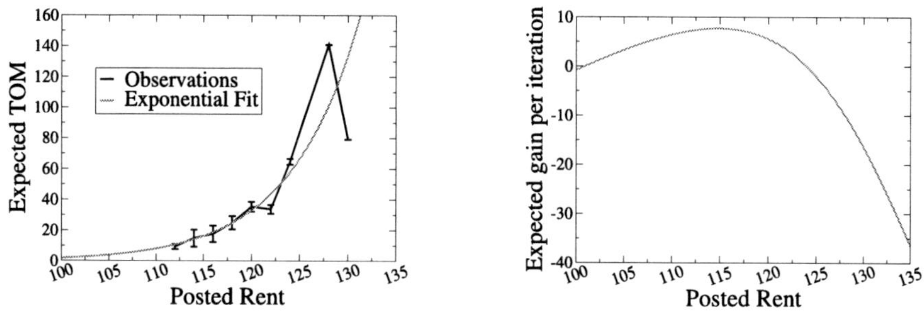

Fig. 2 *Left*: Example of an estimated relation (red line) between asking rent and TOM per rental. The size of the 'error bars' is the statistical weight given to each point in the least-squares fit of the exponential. *Right*: The corresponding expected profits

- A certain fraction $(1/X)$ of tenants, randomly chosen, leave.
- Landlords of newly empty apartments choose their asking rents.
- The next iteration begins.

In Table 1 the parameters of the model are listed, these are used in all results presented, unless stated otherwise. Analysis of the model's behaviour under variations in these parameters can be found in [11]. The default value of X is assumed to be four years (see [15]), meaning that F is three months.

3 Results

As can be seen in Fig. 3-*Left*, the basic model converges for any initialization to a reasonable steady-state with a positive vacancy rate, rent dispersion and nonzero search times.

Table 1 Parameters of the Model

Symbol	Meaning	Default Value of Parameters
	Landlords' parameters	
C_L	Maintenance cost	100
S_L	% of sales seen	20%
F	Timescale rent changes (and memory) (iterations)	15
R_I	Estimation rent interval size	2
	Tenants' parameters	
Y	Housing budget	[100-198]
C_T	Search costs	200
S_T	Percentage of offers seen	5%
X	Expected length of residence (iterations)	240
σ	Idiosyncrasy of tenants preferences (% Y)	5
	General parameters	
size	Town size	10000
Z	Number of initializing iterations	10

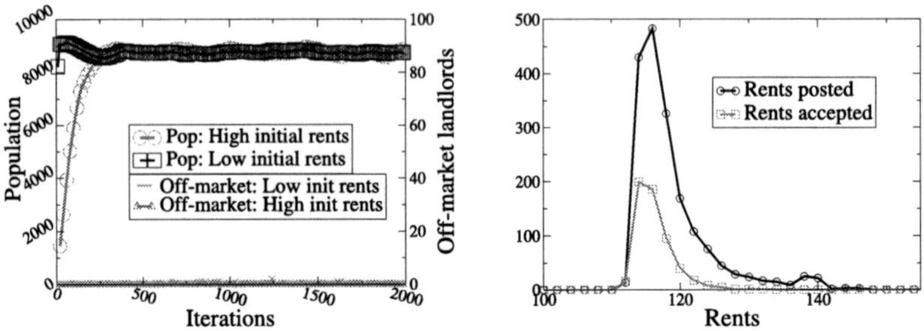

Fig. 3 *Left*: The steady state population and number of landlords off-the-market. *Right*: Rents posted and accepted in last 15 iterations at the steady state, after 2000 iterations

The dispersion in both accepted and posted rents can be seen in Fig. 3-*Right*. We see that most landlords offer rents close to the 'going rate'. The few who ask higher rents are less likely to find tenants. As in existing analytical models, the heterogeneity of tenants' incomes and the presence of market frictions contribute to the dispersion of rents. Additional factors contributing to the dispersion of prices are the idiosyncratic preferences of tenants and stochastic information effects: agents observe different samples of market signals and therefore take different decisions.

3.1 Landlords' Information Level

Landlords' Information Levels, S_L were varied for both homogeneous and heterogeneous landlord populations. Increasing homogeneous landlords' information decreases rents, Fig. 4-*Left*. Landlords need accurate two dimensional information, that is rents offered and their associated times-on-the-market to make good decisions. Their information on TOM for different rents is based on a finite sample accumulated over F iterations. The over-estimations of the optimal rent of less informed

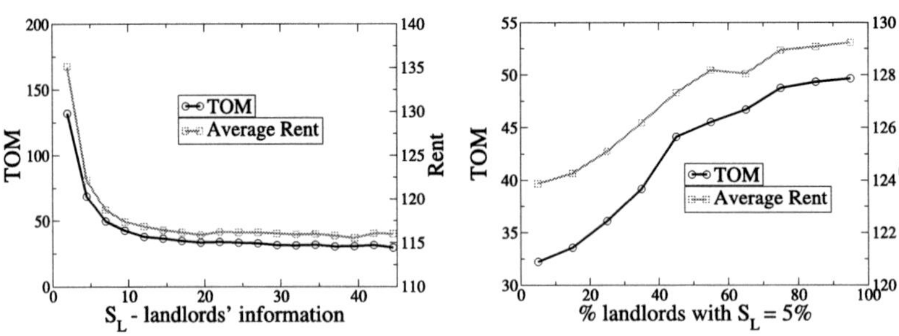

Fig. 4 *Left*: The average rent and the average TOM for homogeneous landlords. *Right*: The average rent and the average TOM as a function of the percentage of landlords who see 5% of the available information while all other landlords see 20%

landlords result from their underestimation of TOM. Ill-informed landlords are less likely to see the long waiting times (TOM) for very high rents which are rarely accepted. This leads to higher asking rents. As the landlords are homogeneous, and make the same errors on average, this pushes the market rents upwards. Every high posted rent, if refused by tenants, increases their expected search times as tenants' search is undirected. This necessarily affects searchers' optimal reservation utilities, pushing the market towards higher rents. In contrast, increasing landlords' information makes them sharper competitors, leading to reduced rents.

In order to further test the effect of landlords' information on the steady-state of the market we perform simulations with landlords who are *heterogeneous* in information. There are two types of landlord, those with the default level of information $S_L = 20$ and those with $S_L = 5$, values that were chosen because Fig. 4-*Left* shows that the steady state changes significantly between these two values when they are shared by all landlords. Figure 4-*Right* shows the increases in average rents and times-on-the-market with the proportion of ill-informed landlords. This is because errors made by ill-informed landlords tend to lead to higher rents. This changes the distribution seen by tenants who have no option but to lower their reservation utilities. As a consequence, well-informed landlords react to the reduced TOM for a given rent by increasing their asking rents as well.

In Fig. 5 we see that, due to the increase in rents, the welfare of both types of landlords increases as the proportion of ill-informed landlords rises. However, the better informed always have higher welfares. This results from their more accurate appreciation of the state of the market. In summary, there are positive externalities (or, more precisely, market effects) of ill-informed on well-informed landlords, the former moving the market rent upwards.

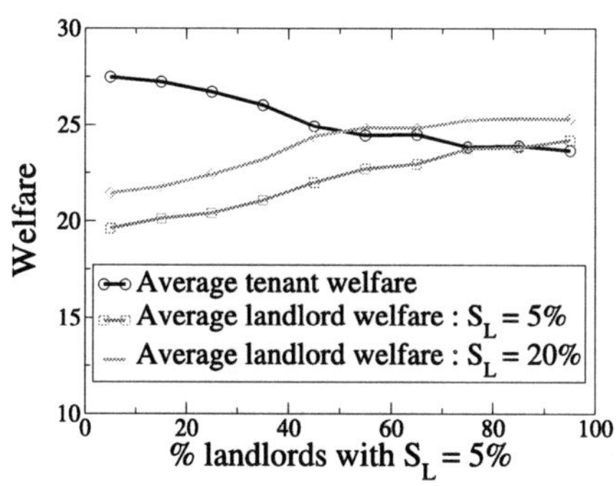

Fig. 5 The average welfares (current utility flows) of tenants and both types of landlord as a function of the percentage of landlords who see 5% of the available information while all other landlords see 20%

3.2 Dynamically Varying the Discount Rate

The discount rate incorporates the value of time, represented by the real interest rate.[3] The discount rate was varied from less than 1% per annum to over 17% per annum. Increasing the discount rate means an increase in the impatience of all agents. That is, at constant rents and TOM, for both categories of agent the value of a match increases with respect to the value of continuing to search. More specifically, changing the discount rate alters all four terms in the right-hand sides of Eqs. 1 and 2.

There ensues two contradictory effects: Landlords have a tendency to post lower rents, while tenants are willing to accept higher rents, conditionally on their income. It is not obvious which of these effects should dominate. In general, one may anticipate that the effects of changes in the discount rate depend on the relative influence of tenants and landlords in the market. As landlords post prices which cannot be negotiated, while tenants decide whether or not to accept the offer received, we can expect landlords' decisions to lead the market.

Figure 6-*Left* shows that for the default values of the other parameters, the average rent is lower with a higher discount rate. This shows that changes in landlords' behaviour, due to a change in discount rate, have a greater impact on market outcomes than the corresponding changes in tenants' behaviour. Increases in the discount rate also lead to a reduction in TOM and an increase in population. Therefore, the average welfare of tenants is improved and that of landlords disimproved with increasing discount rates, as can be seen in Fig. 6-*Right* .

The role of information in the context of a dynamic evolution of real estate markets is an important subject that has already been analysed empirically by [7, 5]. In order to examine this question we introduce exogenous shocks to the discount rate. Two aspects of information play an important role in the decisions of landlords: Firstly the proportion of available information seen and secondly the length of their memory. At the steady state, these two parameters have an equivalent role: both increase the quantity (and therefore quality) of information. However, in an evolving

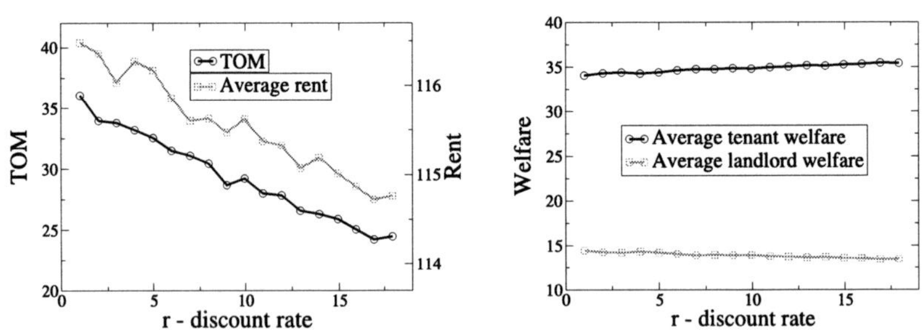

Fig. 6 *Left*:The average rent and the average time-on-the-market for residences rented over the last 15 iterations. *Right*: The average welfares of tenant and landlord agents. Both graphs for a variation in the annual discount rate of both agent types

[3] Note that for our agents only the current discount rate plays a role in their decision.

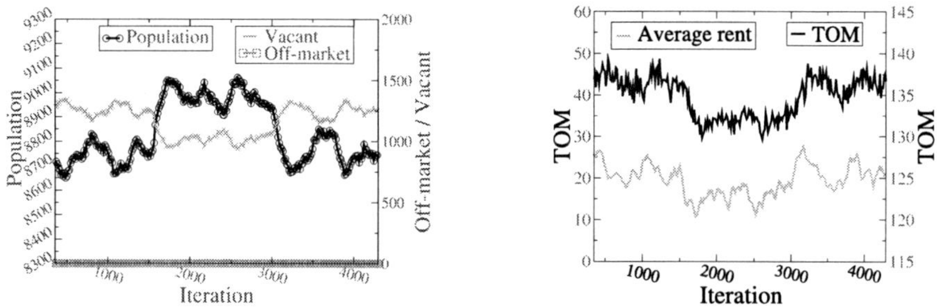

Fig. 7 *Left*: The variations in population, vacancies and the number of landlords off the market, with the varying discount rate r shown *Right*. *Right*: The exogenous variation of the discount rate and the corresponding average TOM

market, the length of memory becomes a two-edged sword. It increases the quantity of information but much of this information may be out-of-date.

The discount rate was varied from 2% to 10% at 1500 iterations and reduced again to 2% at 3000 iterations. The adjustments in the rent and the TOM due to the changes in the discount rate can be seen in Fig. 7-*Right.* As expected from the comparative statics shown in the previous subsection, both the average rent and TOM reduce after the increase in the discount rate. This causes an increase in population due to the larger number of tenants who can afford housing (see Fig. 7-*Left*). Opposing adjustments occur when the discount rate comes back to its previous value. This gives us the opportunity of exploring the dynamics of the market in situations of both rising prices and falling prices.

All landlords who review their rent after the increase at 1500 iterations are aware of the increase in the discount rate, and therefore post lower rents. Their beliefs on the state of the market evolve more slowly due to their memory of past transactions. Therefore, as tenants' reactions to the new situation are not taken into account immediately, accepted rents reduce abruptly, see Fig. 8-*Left.*

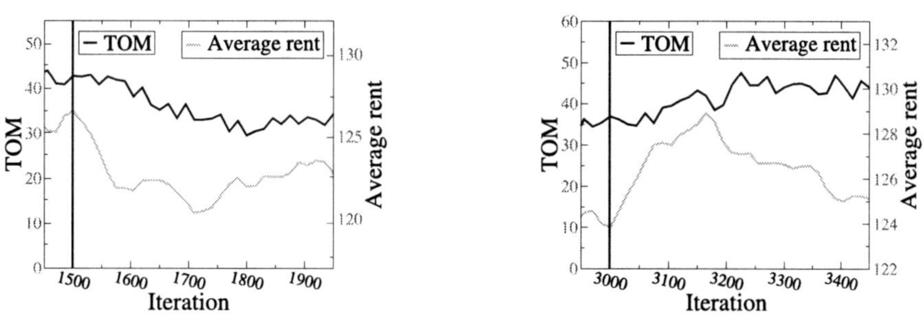

Fig. 8 *Left*: The variations in TOM and average rents, around the transition of the discount rate from $r = 2\%$ to $r = 10\%$ at 1000 iterations. *Right*: The variations in TOM and average rents, around the transition of the discount rate from $r = 10\%$ to $r = 2\%$ at 2000 iterations

However, the average TOM of agreed rents reduces considerably more slowly. Tenants change immediately their reservation utility in reaction to the flow of new low rents. Therefore the average individual acceptance probability reduces due to newly increased value of waiting for these particularly low rents. This hinders the decrease in TOM that would otherwise result from the greater number of tenant agents who can afford housing. Once the rent distribution stabilises, that is the newly posted rents cease to be cheaper than those on the market, average individual acceptation probabilities increase. We observe in Fig. 8-*Left* that just before the average TOM reaches its new steady-state value the rent is at its lowest level. Coupled with the greater number of tenant agents who can now afford housing, this causes the volume and population to rise until the new steady-state is reached. Figure 9-*Left* shows that the population rises by approximately 250 in 200 iterations, equivalent to about $2\frac{1}{2}$ years. As the number of departures is a constant fraction of the population, at the steady-state the volume of transactions is directly related to the population. While the population is rising the volume of transactions is greater than at a steady-state with the same population. Once the rents have ceased to reduce the volume of transactions increases for a short period, around 1570 iterations, as seen by the sharp rise in the population in Fig. 9-*Left*. In fact the volume of transactions increases by over 15% temporarily before lowering to its steady state value that, like the new population, is approximately 3% above its previous steady-state value. The higher volume of transactions in conjunction with a smaller number of vacancies at the new steady-state keep TOM low.[4] Changes in the ratio of transaction volumes to vacancies are an essential element that differentiate 'hot' and 'cold' markets.

After the reduction in the discount rate at 3000 iterations opposite adjustments are seen: an increase in average rent, TOM and the vacancy rate with a decrease in population. A decrease in volume follows from the falling population seen in Fig. 9-*Right*. The rent increases immediately as landlords are instantly informed of the reduction in the discount rate, see Fig. 8-*Right*. What's new here is first that the

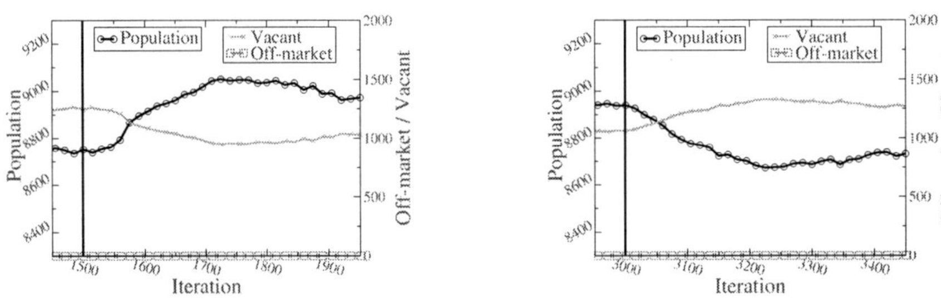

Fig. 9 *Left*: The variations in population, vacancies and the number of landlords off the market, around the transition of the discount rate from $r = 2\%$ to $r = 10\%$ at 1000 iterations. *Right*: The variations in population,vacancies and the number of landlords off the market, around the transition of the discount rate from $r = 10\%$ to $r = 2\%$ at 2000 iterations

[4] The average TOM for landlords is proportional to the number of vacancies divided by the volume of transactions in an iteration. Here, the change in TOM is primarily due to the change in the number of vacancies, which changes by around 20% rather than to the 3% change in volume.

population falls immediately with the increase in rent as poorer tenants have a hard income constraint, see Fig. 9-*Right*.

Secondly, there is a marked overshoot in rents, see Fig. 8-*Right*. This can be attributed to the fact that the eventual negative effects of asking excessive rents take time to be understood by landlords. This is due to both the low frequency of acceptation of high rents, which means that they are often unobserved by individual landlords, and secondarily the relatively long time required for these durations to happen.

4 Conclusion

Our dynamic model includes imperfect information and heterogeneous interacting agents. It leads to price dispersion, nonzero search times and vacancies, three essential ingredients of any realistic housing model. The matching probability depends endogenously on the posted price of apartments.

The heuristics of real world landlords are simulated here by a regression and profit calculation, with a larger number of individual information points than real agents normally know. In our model landlords set rents which tenant agents accept or refuse. Greater information for landlords disimproves their overall utility due to greater competition. However, when landlords with different levels of information are present on the market, the better informed are consistently better off.

We have examined the comparative static and dynamic effects of a change in the discount rate. Landlords have greater market power as they set the rents among which agents choose. It has been shown that rents are lower with higher discount rates, as landlords cost of search out-weighs the gain from higher rents. There is evidence of overshooting in the adjustment of rents after shocks.

Our main aim has been to construct a model that allows hypotheses on the functioning of the urban rental market to be investigated. We believe that a dynamic model based on straightforward micro-economic behaviours with imperfect information is a good approach. We have found robust and simple agent dynamics (or rules) that reproduce important features of the rental housing market.

The current set-up allows the investigation of the distributive effects of policy decisions among tenant agents of varying incomes. Rent control is one possible example [3], as is the level of information among tenants [4].

The agent-based approach adopted here allows many sources of heterogeneities, that cannot be modelled analytically, to be included. It also has considerable potential for modelling the dynamics of housing markets.

Acknowledgements We thank Nicolas Coulombel for useful comments on an earlier draft.

5 Appendix: Initialisation

The landlords all have an initial asking rent randomly chosen in the interval 100-120. The tenant agents have a uniform distribution of housing budgets between 100 and 198 in 50 discrete groups. Over the first Z iterations, tenant agents see five apartments and select the lowest asking rent if it offers the agent a positive utility. This preference for lower rent residences initialises the market in such a way that the information available to landlords indicates that higher rents mean longer waiting times. Landlords do not review their rents during the initialisation phase. After the Z initialisation iterations are complete, the mechanism described in the body of the text is implemented, in which searchers see only one residence per iteration.

References

1. Allen, M., Rutherford, R., T.A.(2009) Residential asking rents and time on the market. Journal of Real Estate Finance and Economics **38**(4):351–365
2. Arnott, R. (1989) Housing vacancies, thin markets, and idiosyncratic tastes. Journal of Real Estate Finance and Economics **2**:5–30
3. Bradburd, R., Sheppard, S., Bergeron, J., Engler, E. (2006) The impact of rent controls in non-walrasian markets: An agent-based modeling approach. Journal of Regional Science **46**(3):455–491
4. Bradburd, R., Sheppard, S., Bergeron, J., Engler, E., Gee, E. (2005) The distributional impact of housing discrimination in a non-walrasian setting. Journal of Housing Economics **14**(2):61–91
5. Clayton, J., MacKinnon, G., Peng, L. (2008) Time variation of liquidity in the private real estate market: An empirical investigation. Journal of Real Estate Research **30**(2):125–160
6. Desgranges, G., Wasmer, E. (2000) Appariements sur le marche du logement. Annales d'economie et de statistique (58):253–287
7. Fisher, J., Gatzlaff, D., Geltner, D., Haurin, D. (2003) Controlling for the impact of variable liquidity in commercial real estate price indices. Real Estate Economics **31**(2):269–303
8. Gabriel, S., Nothaft, F. (1988) Rental housing markets and the natural vacancy rate. American Real Estate and Urban Economics Association Journal **16**(4):419–429
9. Grenadier, S. (1995) Local And National Determinants Of Office Vacancies. Journal Of Urban Economics **37**(1):57–71
10. Hwang, M., Quigley, J. (2006) Economic fundamentals in local housing markets: Evidence from us metropolitan regions. Journal of Regional Science **46**(3):425–453
11. Mc Breen, J., Goffette-Nagot, F., Jensen, P. (2009) An agent-based simulation of rental housing markets. `ftp://ftp.gate.cnrs.fr/RePEc/2009/0908.pdf`
12. Rogerson, R., Shimer, R., Wright, R. (2005) Search theoretic models of the labour market: A survey. Journal of Economic Literature **63**:959–988
13. Rosen, K., Smith, L. (1983) The price-adjustment process for rental housing and the natural vacancy rate. American Economic Review **73**(4):779–786
14. Shilling, J., Sirmans, C., Corgel, J. (1987) Price adjustment process for rental office space. Journal of Urban Economics **22**(1):90–100
15. de Una-Alvarez, J., Arevalo-Tome, R., Soledad Otero-Giraldez, M. (2009) Nonparametric Estimation of Households' Duration of Residence from Panel Data. Journal Of Real Estate Finance And Economics **39**(1):58–73, doi:10.1007/s11146-007-9102-2
16. Wheaton, W. (1990) Vacancy, search, and prices in a housing market matching model. Journal of Political Economy **98**(6):1270–1292

Adaptation of Investments in the Pharmaceutical Industry

Tino Schütte

Abstract Situated in the research field of market structure and strategic behavior, a model of product market competition is developed, showing the impacts of investment adjustments on the success of companies. Placed in an agent-based multi-firm multi-product setting, we study the consequences of the two-step decision of firms to split their budgets in (i) marketing and development activities and (ii) development expenditures into innovative or imitative activities. The model is validated with empirical data of the pharmaceutical industry, with reference to the drug market in Germany. The results show that investment strategies adjusted to the behavior of direct competitors outperforms adjustments based on individual aspiration levels.

1 Introduction

Investments are one of the most important business processes particularly for the development of new (exploration) and the utilization of existing market potential (exploitation). They are necessary to improve turnovers and to survive competition in the long run. As market conditions on non-stationary markets change perpetually, an ongoing decision making about investment adjustments is inevitable. Special relevance has the adaptation of action parameters that influence product differentiation and thus can enhance the firm's performance compared to its competitors.

Previously there is no study with an industrial economic perspective that explicitly investigates the impact of alternative investment adjustment strategies on firms in a competitive product market. Therefore the following research question will be answered: How do investment adjustments contribute to the success of firms operating in a market that is primarily characterized by product competition? The task of the paper is therefore (1.) the modeling of a representative product market

Tino Schütte
Technische Universität Dresden, Faculty of Economics, 01062 Dresden
e-mail: tino.schuette@tu-dresden.de

M. LiCalzi et al. (eds.), *Progress in Artificial Economics*, Lecture Notes in Economics and Mathematical Systems 645, DOI 10.1007/978-3-642-13947-5_14,

competition and (2.) the analysis of the consequences of alternative investment strategies for firms in that market. Precise statements about changes in business success by using alternative adaptation scenarios will be given and recommendations for future investment decisions will be derived. To validate the model empirically an existing market with typical characteristics of product market competition is investigated, the drug market. The pharmaceutical industry is considered, because it is active in a market described by intense innovation and imitation competition. Based on stylized facts of the pharmaceutical industry (as the basis of modeling), the simulation model is calibrated with empirical data of the drug market in Germany[1].

To answer the research question the method of agent-based modeling and simulation (ABMS) is used.[2] The computer simulation allows repeatable experiments and thus a detailed counterfactual analysis of alternative investment adjustment strategies. The investment strategies are implemented in the form of well defined conditional-action-rules on the agent level and their impact on firms' performance is examined. To evaluate the simulation results, standard statistical methods, in particular analysis of variance, is used. The focus of the analysis is set on the development of firms' sales interpreted as the available budget for investments in the following period.

The structure of the paper is as follows. After shortly classifying the pharmaceutical industry, a formal model is developed, which is implemented using NetLogo 4.0. After verifying the source code and validating the programmed simulation model, a reference simulation is established, which gives the basis for the consequent simulation analysis.

2 Modeling product market competition

According to [5], the pharmaceutical industry can be stylized as a 'Type2AR'-Industry. That means that especially vertical product differentiation through advertisement (A) and research and development (R) is essential. The market is characterized by intense product competition. Despite the fact that the drug market is quite fragmented, two major segments can be distinguished: (i) originals and 'me-too' drugs (patent protected) as well as (ii) generics[3].

---

[1] It is not the aim to represent the the development of the German market as detailed as possible but to represent important principles of a competitive product market in a load-bearing simulation model.

[2] Advantages of using ABMS are e.g.: taking microeconomic perspective of modeling, using agents with bounded rational behavior, considering strategic interactions, endogenize novelties, allowing self organizational and emergent structure. For a further discussion of the appropriateness of ABMS see e.g. [3], [7] or [6].

[3] Because of the different areas of indication the substitutability of products (drugs) is low. The drug market is therefore a market with numerous submarkets. Within these submarkets there nevertheless can be a high market concentration. For an analysis of the pharmaceutical industry see e.g. [2].

The stylized facts of a Type2AR-Industry serve as a starting point for a formal model that is capable of representing fundamental principles of the product market competition in the pharmaceutical industry. These include in particular the conflicts between research and development vs. marketing (action parameter ε) and between innovation vs. imitation (action parameter α). The initial point of consideration are the firms product portfolio. They offer products with and without patent protection. Firms are seen as individuals with their own behavior and who can independently make decisions. According to their development priority they will be assigned to innovators ($\alpha_i \geq 0,5$) or imitators ($\alpha_i < 0,5$).

2.1 Modeling the Supply Side

In each period t a firm i must decide between the proportion of the available budget B_i that is invested in the marketing V_i ('exploitation' of the existing product portfolio) and the proportion that is invested in the development E_i of products ('exploration' to expand the product portfolio). The decision is made on the basis of determining the budget allocation parameter $\varepsilon_i \in [0,1]$. The allocation of the budget $B_{i,t}$ is thus described by

$$B_{i,t} = E_{i,t} + V_{i,t} = \varepsilon_i B_{i,t} + (1-\varepsilon_i) B_{i,t}. \tag{1}$$

In addition to that, a firm has to decide between the proportion of development expenditure it spends on product innovation and the part it wants to invest in product imitations. The development expenditure for innovations is described by the proportion $\alpha_i \in [0,1]$. The remaining part of $(1-\alpha_i)$ stands for the proportion available for imitations. The decision tree of a firm i ($i = 1,...,n$) can therefore be described as a two stage decision tree as shown in Fig. 1. It should be mentioned that production costs are ignored, because this paper focuses on the reflection of the heterogeneity in competition. Production costs of drugs are comparable for all firms.

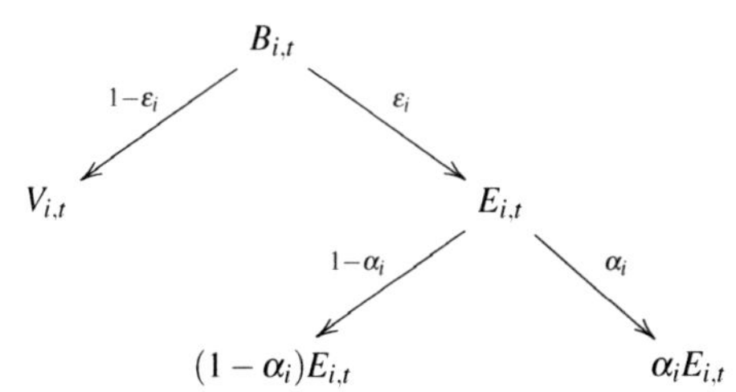

Fig. 1 Decision tree of a firm i in period t

Firms compete with each other by allowing them to bring new products to the market and imitate existing products. A new product generates a new demand, which is totally assigned to the patentee due to the available patent protection. After the expiry of the patent, other companies can offer the product as well and participate in the sales volume. Products that do not generate a minimum revenue, disappear from the market. Taking into account the intensity of competition and the marketing efforts, each firm can get a part of the total product demand.

Marketing expenditures V and investments in product imitation $(1-\alpha)E$ are investments 'without memory' and therefore only relevant in the current period. Investments in product research and development αE however will help to develop a knowledge stock W. This is the crucial variable for the probability of bringing a new product to the market.[4] Taking into account a half-life of knowledge ξ and a research productivity η, knowledge can be accumulated over time.

$$W_{i,t} = \xi W_{i,t-1} + \eta \sqrt{\alpha_i E_{i,t}} \tag{2}$$

In the model, a firm can launch an innovation if its knowledge stock $W_{i,t}$ exceeds a randomly generated innovation threshold $S_{i,t}^{Inno}$. This threshold may differ for each firm from period to period. If a firm implements an innovation in the form of a patented product, its knowledge stock is reset to 0, because it is assumed that knowledge is product specific and has no relevance for future innovation.

$$Innovation: \begin{cases} yes, & \text{if } W_{i,t} > S_{i,t}^{Inno} \sim N(\mu_S, \sigma_S) \\ no, & \text{if } W_{i,t} \leq S_{i,t}^{Inno} \sim N(\mu_S, \sigma_S) \end{cases} \tag{3}$$

An innovation is patented for T^{Pat} periods so it can't be imitated by other firms during that time. After patent expiry other firms can imitate the product. For an imitation, costs of a fixed size S^{Imi} are required. This condition describes the necessary efforts and authorization of a drug before being merchandised (regardless of whether it is patented or not).

$$Imitation: \begin{cases} yes, & \text{if } (1-\alpha_i)E_{i,t} > S^{Imi} \\ no, & \text{if } (1-\alpha_i)E_{i,t} \leq S^{Imi} \end{cases} \tag{4}$$

If the firm fulfills the imitation condition, it selects a product without patent protection, which it also wants to produce. The selection of the product is based on

$$max_{j=1,\ldots,m_t} \left(\frac{Q_{j,t}}{n_{j,t}+1} \right). \tag{5}$$

The ratio of current demand $Q_{j,t}$ left on the product $j \in \{1,\ldots,m_t\}$ and the number $n_{j,t}$ of firms that manufacture this product, plus the imitative enterprises, is evaluated. This takes into account the observation that firms tend to imitate products with the largest possible product demand for themselves.

Through innovation and imitation firms augment their product portfolio which is compared each period with the product portfolios of its competitors. Weighted with the relative marketing expenditure, the product demand for each firm is determined.

[4] For a review of the literature on approaches to model innovation and imitation see [4], p.128 ff.

Multiplied with the prices of the products the sales of the current period are calculated. In the subsequent period, this is the available budget. If a firm cannot generate a minimum budget, it will fall out of the product market competition.

$$Market\ exit\ condition: \qquad B_{i,t} < B^{min} \tag{6}$$

This assumption ensures that the model considers only companies that have a sufficient liquidity for investments.

2.2 Modeling the Demand Side

Given the nature of the drug market, the demand side is separated into consumers (the patients), decision makers (the physicians) and buyers (the insurances). Consequently no systematic price sales function can be derived, which describes a functional relationship between price and sales volume. With regard to the quasi price inelastic demand of consumers, the following function, based on an approach of [1], is used.

$$Q_{j,t}(Y,\kappa,t^{Alt}) = \frac{Y}{\ln(\kappa)+1}(\kappa^{t_j^{Alt}} - e^{-t_j^{Alt}}), \tag{7}$$

where Y stands for the potential market volume of a drug, the parameter κ describes the proportion of loyal customers and the age of the drug is denoted by t_j^{Alt}. If the demand for a product falls below a minimum value of Q^{min} it falls out of the market.

$$Market\ exit\ condition: \qquad Q_{j,t} < Q^{min} \tag{8}$$

This assumption helps to consider only products with a minimum of demand.

When a drug is produced by several firms, its demand is divided in accordance with the relative marketing expenditure of the firm in the current period.

$$Q_{i,j,t} = \frac{V_{i,t}}{\sum_{l=1}^{n_t} V_{l,t}} Q_{j,t} \tag{9}$$

The overall demand for a firm is equal to the sum of demands for its products.

$$Q_{i,t} = \sum_{j=1}^{m_t} Q_{i,j,t} \tag{10}$$

The total demand of a period is composed of the sum of all product demands.

$$Q_t = \sum_{j=1}^{m_t} \sum_{i=1}^{n_t} Q_{i,j,t} \tag{11}$$

A further attribute of the products is their price. Each product has a price P_j^{Pat} for the period T^{Pat} when it is patented and a price P^{Gen} for the subsequent periods.

The prices of drugs which are patent protected are modeled as normally distributed random numbers $P_j^{Pat} \sim N(\mu_P, \sigma_P)$. For the price of P^{Gen} an empirical average is used.

$$P_j = \begin{cases} P_j^{Pat} & \text{if } t_j^{Alt} \leq T^{Pat} \\ P^{Gen} & \text{if } t_j^{Alt} > T^{Pat} \end{cases}$$

It is also true that $P_j^{Pat} > P^{Gen}$. Multiplying the product demand of a firm with the product pricing, we get the sales for the current period and thus the budget for marketing and development investments in the subsequent period.

$$B_{i,t+1} = \sum_{l=1}^{m_{i,t}} P_{l,t} Q_{i,l,t} \quad (12)$$

2.3 Statistical Variables

To control the simulation results of the model, a selection of important statistical variables is considered. In addition to basic parameters like the number of firms / products and the budgets, various types of describing variables such as the Herfindahl-Hirschman Index and the Instability Index are taken into account. The Herfindahl-Hirschman-Index is defined as

$$H_t = \frac{\sum_{i=1}^{n_t} B_{i,t}^2}{(\sum_{i=1}^{n_t} B_{i,t})^2} = \frac{1}{B_t^2} \sum_{i=1}^{n_t} B_{i,t}^2 = \sum_{i=1}^{n_t} (\frac{B_{i,t}}{B_t})^2 = \sum_{i=1}^{n_t} c_{i,t}^2 \quad (13)$$

with $c_{i,t}$ being the market share of a firm i in period t. The instability index (I) calculates how market shares of the firms change from one period to another.

$$I_t = \sum_{i=1}^{n_t} |c_{i,t} - c_{i,t-1}| \quad (14)$$

In addition to the Herfindahl-Hirschman Index and the Instability Index the growth of the budget g_B and the growth of the market share g_c of a firm (also over several periods) are measured. Moreover the mean and the standard deviation of sales are considered.

2.4 Model Validation

The product competition model is calibrated with empirical data of the pharmaceutical market in Germany. The German drug market was examined for variables that could be used directly in the configuration of the reference simulation. Remaining parameters and variables are set in such a way that a reproduction of important characteristics of the observed competition is shown. The data base is provided mainly through the 'Arzneiverordnungs-Report 2006'[5], especially the section on economic aspects of the German pharmaceutical market in 2005. The resulting reference model was tested for robustness with a comprehensive sensitivity analysis. On the demand side, especially the influence of the loyalty rate of customers as well as the patent term had been evaluated. On the supply side, the effects of variation in research productivity and the innovation as well as imitation threshold were analyzed. The sensitivity analysis shows that the model responds to changes in accordance with theoretical considerations. The controllability of the simulation model allows a realistic analysis of alternative adjustment scenarios. The used configuration for the reference simulation is shown in Table 1. In interpretation that one period is mapping one quarter, the development of the market over 25 years is simulated.

Firms		Products	
$n_{t=0} = 50$	$B_{i,t=0} = 250'$	$m_{t=0} = 500$	$Y = 500$
$\varepsilon_i \sim U(0,1)$	$B^{min} = 5'$	$\kappa = 0.9$	$P_j^{Pat} \sim N(150;100)$
$\alpha_i \sim U(0,1)$	$S_{i,t}^{Inno} \sim N(200;50)$		$P^{Gen} = 30$
$\xi = 0.95$	$S^{Imi} = 20$		$Q^{min} = 5$
$\eta = 0.3$	$T = 100$		$T^{Pat} = 20$

Table 1 Configuration of the reference simulation

3 Simulating Investment Adjustments

The aim of firms in the formulated agent-based model is to achieve the highest possible product sales in the competitive market. To reach this aim they get the ability to adapt their parameters of action over time. The starting point of adjustment decision is the strategic choice of firms (see Fig. 1). An investment decision in terms of a condition-action rule consists of a scale for measuring success (condition: Do I need an adjustment?) and a way to change the behavior (action: How does adaptation takes place?). Two basic classes of investment strategies of adaptation are distinguished.

1. Firms take their own aspiration level for evaluating their success. In case of underperformance, they will adjust their parameters of action toward the level of the best performing firm on the market.
2. Firms take the average sales level of their direct competitors as a comparison. In case of underperformance, they will adjust their parameters of action slightly toward the average parameter value of their competitors.

[5] German pharmaceuticals prescriptions report.

Both classes are divided into further scenarios. From the multitude of possible options, 1- and 2-year decision intervals are taken. Therefore firms can decide every 4 or 8 periods on the adjustment of budget allocation ($v = 4$ and $v = 8$). The basis of assessment is the development of the budget since the last decision. It is assumed that an adjustment can take place only gradually, because investments are part of long-term business strategies. Thus the possibility of a gradual adjustment of the parameters of action at $\Delta = 0.05$ and $\Delta = 0.1$ is considered.

	Condition	Action (parameter adjustment)
1. Orientation at their aspiration level		
Scenario I:	$g_B^{min} = 0.05$	orientated at the best (ε-adjustment)
Scenario II:	$g_B^{min} = 0.1$	orientated at the best (ε-adjustment)
2. Orientation at the success of the direct competitors		
Scenario III:	$\bar{g}_B$	orientated at the average (ε-adjustment)
Scenario IV:	$\bar{g}_B$	orientated at the average (α-adjustment)
Scenario V:	$\bar{g}_B$	orientated at the average (ε- and α-adjustment)

Table 2 Investigated adjustment scenarios

In the first group of scenarios two aspiration levels of firms budget growth ($g_B^{min} = 0.05$ and $g_B^{min} = 0.1$) are differentiated. Because of limited available information on the investment strategy of the best competitor, firms are only capable to adjust their action parameter ε. The action parameter α remains unchanged. This restriction is necessary, because it can not be assumed that firms know the exact layout of the development budgets of their competitors. This information is only available at branch level. In the second group of scenarios it is distinguished whether only one or both parameters of actions can be adjusted. Each adjustment scenario (I-V) is subdivided with the 4 adaptation options ($v = 4$; $v = 8$; $\Delta = 0.05$; $\Delta = 0.1$) so that a total of 20 investment strategies of adaptation are investigated.

First Group of Scenarios: Orientation at Their Aspiration Level

Condition: comparison of the average growth rate of the previous periods v with a desired growth rate.

$$\frac{1}{v}\sum_{l=t-v}^{t} g_{B_{i,l}} < g_B^{min}$$

Action: adaptation of the action parameter in the direction of the parameter value of the most successful firm k.

$$\varepsilon_{i,t+1} = \begin{cases} \varepsilon_{i,t} + \Delta & \text{if } \varepsilon_{i,t} < \varepsilon_{k,t} \\ \varepsilon_{i,t} - \Delta & \text{if } \varepsilon_{i,t} \geq \varepsilon_{k,t} \end{cases}$$

Second Group of Scenarios: Orientation at the Success of the Direct Competitors

Condition: comparison of their average growth rate with the average growth rate of competitors.

$$\frac{1}{v}\sum_{l=t-v}^{t} g_{B_{i,l}} < \frac{1}{n}\frac{1}{v}\sum_{i=1}^{n}\sum_{l=t-v}^{t} g_{B_{i,l}}$$

Action: adaptation of the action parameter in the direction of the mean parameter value of the direct competitors.

$$\varepsilon_{i,t+1} = \begin{cases} \varepsilon_{i,t} + \Delta & \text{if } \varepsilon_{i,t} < \bar{\varepsilon}_t \\ \varepsilon_{i,t} - \Delta & \text{if } \varepsilon_{i,t} \geq \bar{\varepsilon}_t \end{cases}$$

The extensive simulation analysis gives the following summarized data.

3.1 Success of Strategies: Innovative Firms

The values marked with an asterisk are statistically significant differences at the 0.05 level of significance (Welch-Test) compared to the reference simulation (486.927). The data show that the strategies of scenario I and II cannot increase budgets

	Scenario I	Scenario II	Scenario III	Scenario IV	Scenario V	
Strategy	**1-4**	**5-8**	**9-12**	**13-16**	**17-20**	
Parameter	$g_B^{min} = 0.05$	$g_B^{min} = 0.1$	ε	α	ε, α	
$v = 4$						
$\Delta = 0.05$	454.135* (89.223)	440.065* (95.356)	545.337* (38.145)	507.070* (40.598)	554.870* (42.597)	(a)
$\Delta = 0.1$	482.119 (121.349)	474.586 (121.067)	562.894* (40.378)	496.112 (41.607)	576.405* (41.715)	(b)
$v = 8$						
$\Delta = 0.05$	433.988* (99.933)	415.201* (106.007)	534.520* (41.848)	496.214 (44.817)	547.855* (57.436)	(c)
$\Delta = 0.1$	492.695 (102.107)	385.823* (126.068)	546.803* (47.857)	489.914 (44.263)	563.200* (50.943)	(d)

whereas the strategies of scenario III and IV improve budgets. It is also obvious that an adjustment of ε is preferred, compared to an adjustment of α. A visualization of the outcome helps to clarify the trends.

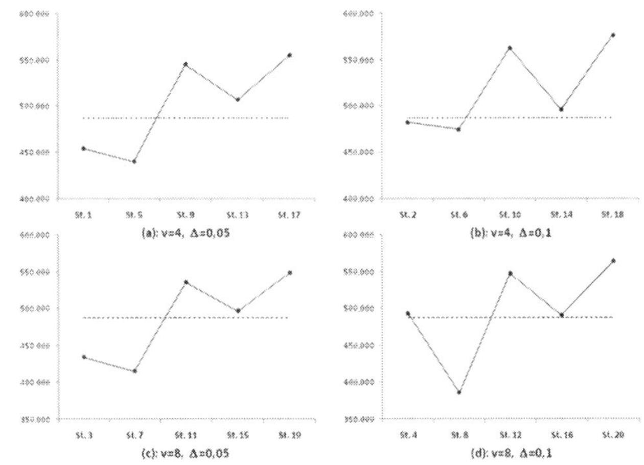

Fig. 2 Development of budget averages: innovative firms

3.2 Success of Strategies: Imitative Firms

The asterisk marked values are significant differences on the 0.05 level of significance (Welch-Test) compared to the reference simulation (295.214). The data show

	Scenario I	**Scenario II**	**Scenario III**	**Scenario IV**	**Scenario V**	
Strategy	**1-4**	**5-8**	**9-12**	**13-16**	**17-20**	
Parameter	$g_B^{min} = 0.05$	$g_B^{min} = 0.1$	ε	α	ε, α	
$v = 4$						
$\Delta = 0.05$	241.420* (52.797)	227.348* (49.094)	285.677 (37.082)	288.073 (42.939)	274.903* (35.467)	(a)
$\Delta = 0.1$	261.261* (74.600)	253.391* (67.125)	296.876 (38.217)	316.035* (58.733)	309.223* (51.547)	(b)
$v = 8$						
$\Delta = 0.05$	231.266* (58.580)	220.881* (57.864)	283.645 (38.741)	282.793* (44.667)	273.125* (31.721)	(c)
$\Delta = 0.1$	264.215* (61.508)	201.564* (66.480)	291.307 (42.871)	311.033* (59.355)	290.631 (41.760)	(d)

that the strategies of scenario I and II cannot increase budgets whereas the strategies

14, 16 and 18 improve budgets. An adjustment of α is preferred compared to an adjustment of ε!

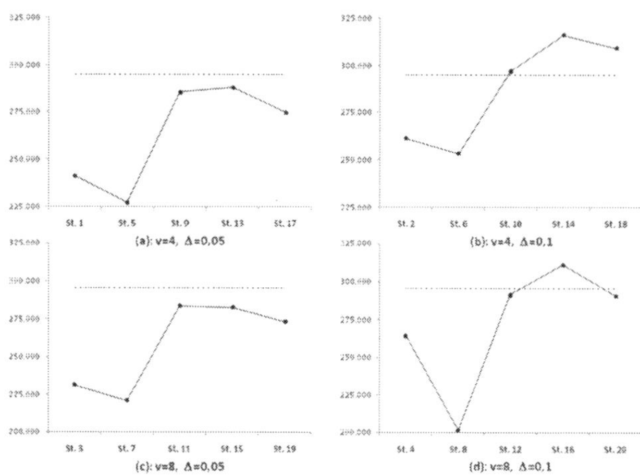

Fig. 3 Development of budget averages: imitative firms

4 Conclusions

The following results (based on predefined hypotheses) are most important.

1. The average budgets can only be increased by some investment adjustment strategies. A focus on the success of direct competitors is preferable to a focus at a fixed aspiration level.
2. An extension of decision intervals decreases budgets whereas an increased scope of adjustment usually improves the budget development. The strategies with $v = 4$ and $\Delta = 0.1$ gave usually the best results within a scenario.
3. A higher aspiration level does not improve budget development necessarily. The results of scenario II are worse than the ones of scenario I.
4. For innovative firms an adjustment of ε is more efficient than an adjustment of α. For imitative firms the opposite is true.

5. For innovative firms a simultaneous adjustment of both parameters is beneficial. For imitative firms the separate adjustment of α is beneficial.
6. Innovative firms gain more from investments adjustments than imitative firms.

The investigation has shown that firms should ground their adaptation strategy not only on a fixed aspiration level but also on their relative position compared to their direct competitors. Moreover the selection of the parameters of action for investment adjustments is critical. Firms should choose their own adaptation strategy depending on the type and their aims. It is not per se advantageous to change as many action parameters as possible. The relevant parameters have to be identified and be altered in compliance with the feasible scope for adjustment and adaptation intervals.

References

1. Debenham, J., Wilkinson, I. (2006) Exploitation versus exploration in market competition. Industry and Innovation **13**:263–289
2. Fischer, D., Breitenbach, J. (2007), Die Pharmaindustrie - Einblick, Durchblick, Perspektiven. Spektrum Akademischer Verlag
3. Gilbert, N., Troitzsch, K.G. (2005) Simulation for the Social Scientist. Open University Press
4. Lehmann-Waffenschmidt, B.C. (2006) Industrieevolution und die New Economy. Metropolis
5. Matraves, C. (1999) Market structure, R&D and advertising in the pharmaceutical industry. Industrial Economics **47**:169–194
6. North, M.J., Macal, C.M. (2007) Managing Business Complexity. Oxford University Press
7. Tesfatsion, L. (2006) Agent-based computational economics. In: Handbook of Computational Economics 2. North-Holland

An Agent-Based Information Management Model of the Chinese Pig Sector

Sjoukje A. Osinga, Mark R. Kramer, Gert Jan Hofstede, Omid Roozmand, and Adrie J.M. Beulens

Abstract This paper investigates the effect of a selected top-down measure (what-if scenario) on actual agent behaviour and total system behaviour by means of an agent-based simulation model, when agents' behaviour cannot fully be managed because the agents are autonomous. The Chinese pork sector serves as case. A multi-level perspective is adopted: the top-down information management measures for improving pork quality, the variation in individual farmer behaviour, and the interaction structures with supply chain partners, governmental representatives and peer farmers. To improve quality, farmers need information, which they can obtain from peers, suppliers and government. Satisfaction or dissatisfaction with their personal situation initiates change of behaviour. Aspects of personality and culture affect the agents' evaluations, decisions and actions. Results indicate that both incentive (demand) and the possibility to move (quality level within reach) on farmer level are requirements for an increase of total system quality. A more informative governmental representative enhances this effect.

1 Introduction and Background Literature

Often, there is a discrepancy between the desired effect of a policy measure and its actual effect, which is a result of failure to account for the behaviour of the target population. A recent example from the Chinese pork sector was reported in the China Daily of January 7, 2010 [3]. Because of pork quality and safety reasons, government officials in Gaobu town of Dongguan in Guangdong province contracted the supply of live pigs to one particular wholesaler, under the impression that this would not affect the general market order. As a result, over 60 pork shops in the largest market in town for fresh pork were closed due to the burden of increased

Wageningen University, P.O. Box 8130, NL-6700 EW Wageningen, The Netherlands
e-mail: {Sjoukje.Osinga,Mark.Kramer,Gertjan.Hofstede,Omid.Roozmand, Adrie.Beulens}@wur.nl

M. LiCalzi et al. (eds.), *Progress in Artificial Economics*, Lecture Notes in Economics and Mathematical Systems 645, DOI 10.1007/978-3-642-13947-5_15,

costs. Since pork is a major food in China, especially during the holiday season, this caused a strong dissatisfaction among the town residents.

The objective of this study is to gain insight in the relationship between sector level information management strategies and actual behaviour of individuals. This insight is obtained by means of agent-based modelling, because this allows us to perform simulation experiments on the behaviour of individuals which are impractical or impossible in the real world. The objective can only be met by applying this general problem to a suitable realistic case. In the Chinese pork sector the government has targets and requirements, but the majority of producers consists of individual farmers who act on their own authority. A research assumption is that in order to reach the government's targets or to fulfil their requirements, farmers need to have certain information. The government can make an effort to disseminate this information, but whether the farmers receive it and act upon it as well, is something beyond control of the government.

Up to now, information management theory has applied a normative approach: information management models usually depict a priori designed flows of tasks, procedures and responsibilities. However, little research has been done to measure the actual effectiveness of applying such models [10]. Research does indicate that there is a gap between the high-level models and the actual behaviour of individuals: information in social and professional networks does not only travel along the lines of formal models [1]. This is especially true in rural communities [15]. Information management as a research field is currently in need of models that integrate actual behaviour with prescriptive models [6]. Such models should allow system level interventions as well as include behaviour of individual actors who are part of the system, where much also depends on the interaction structure between individual actors. This boils down to a multi-level system view.

Agent-based modelling (ABM) is becoming popular in the social sciences and also in information management because it allows representing individual behaviour as a conjunction of reasoning (decision making), personality and values [9]. Focussing on model purpose, Gilbert divides multi-agent models into facsimile models, abstract models and middle range models [9] (p.40 et seq.). Middle-range models aim to "describe the characteristics of a particular social phenomenon, but in a sufficiently general way that their conclusions can be applied widely". Gilbert introduces the aspect of qualitative resemblance. Moss [18] specifies this as: "The dynamics of the model should be similar to the observed dynamics, and the results of the simulation should reveal the same or similar 'statistical signatures' as observed in the real world; that is, the distributions of outcomes should be similar in shape". Our conclusion is that the strong points of ABM match the requirements of our study, and that a middle-range model fits best to integrate a multi-level view in an information management domain.

As a case study, we model Chinese pig farmers who run a family business and earn a living out of pig farming. The majority of Chinese pig production, which adds up to 50% of all pork in the world, comes from small-scale farms with up to a few hundred animals [7]. The case is significant because it helps to gain insight into opportunities for this sector to enhance product quality by means of improved information management strategies.

Furthermore, the choice for this case study was made because it has characteristics that are very attractive for a multi-level agent-based simulation model. Most prominent is the fact that multiple levels are a characteristic of Chinese society: unlike the Netherlands, China's centralized government has the power to implement measures in a relatively short time, in a vertical chain through successively lower levels of government. Responsibilities are person-based rather than rule-based within a multi-layered hierarchical structure [16]. The government has clear targets in these times of economic growth: both pork volume and quality must increase [2]. It would be desirable for the government to be able to address all farmers at once, but this is impossible. To visit every single farmer is not cost-effective, and technological means (like computers and internet) are insufficiently available for any chance of success. The population of farmers is heterogeneous: any strategy implies assumptions about whether the information will actually reach a particular farmer and whether he will adopt the advice [19]. Much depends on farmers' social network [17], personal situation, personality [4] and values [12].

Finally, behaviour is not independent: like fashion, it can spread within a population. If one farmer gains profit from his decision (in money, or in reputation), he will do it again, and it is likely that others will follow his example. Such influences add extra dynamics to a population, and even cause sweeping changes in behaviour. The hog cycle [11] is an iconic example in this respect. The aspect of feedback/feedforward loops and the emerging behaviour make the case very interesting for the ABM research community.

2 Problem Definition

The specific focus of this paper will be on the representation power of agent-based modelling when applied to the Chinese pig farmer case, and to simulate the effect of a selected top down policy measure (what-if scenario) on actual agent behaviour and total system behaviour. The research questions are: Can we adequately represent the real-world case, and: can we implement the selected scenario and how plausible are the effect(s) that we find in reality?

Figure 1 shows our conceptual model from focal farmer perspective. We apply three modelling levels (system, agent and interaction level) as described in [5]. System level characteristics include informing behaviour of the governmental Livestock Bureau Official (LBO) and the availability of suitable supply chain and network (SCN) partners, i.e. pig buyers, and other farmers (friends). Interaction level characteristics entail opportunities to meet business partners and friends. Agent level characteristics include personality characteristics (motivations and abilities) that influence the actual application of acquired information [20].

We assume that more information leads to higher quality pigs, provided that (a) the information reached the farmer and (b) the farmer chose to apply the knowledge.

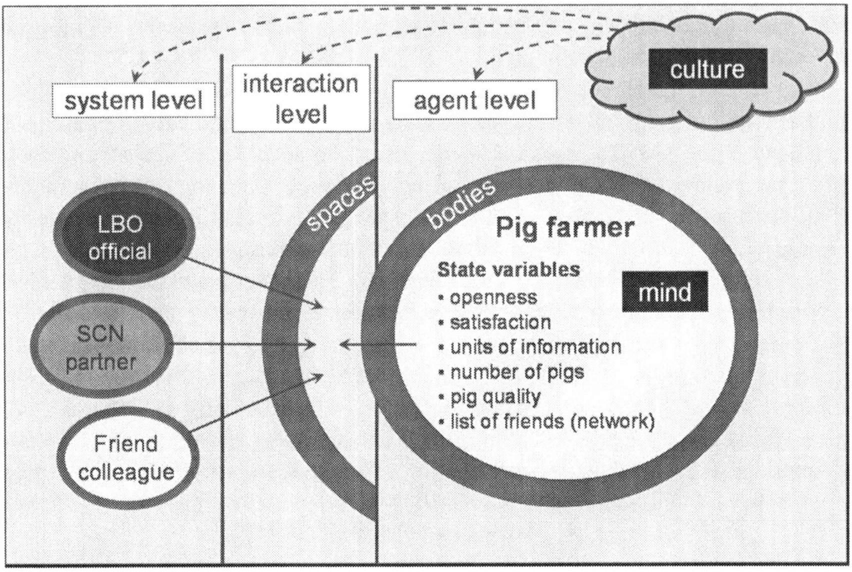

Fig. 1 Conceptual model for our agent-based model from viewpoint of focal pig farmer, with multiple levels (system, agent, interaction). References to MASQ-architecture (mind, bodies, spaces, culture) are explained in Sect. 3

We also assume that a farmer with a social network can share his friends' knowledge more easily than a farmer who has no such connections.

For the sector as a whole, representation of farmers in all market segments is desirable. The best system behaviour is not simply to attain a total pig quality as high as possible. Demand varies with respect to specific products and also depends on marketing channel. Certain pork products require high quality pig meat (e.g. cutlets for restaurants), but it is more practical to use lower quality for other pork products (e.g. sausages for the fresh market). Ideally, there is a balance between supply and demand for certain quality. When demand changes—a global, system level change—farmers should take decisions as well—at local level—in order to adapt.

The system level intervention we select for our what-if experiment is to diminish the demand for low quality pigs, inspired by the news article we used in the introduction of this paper. Based on the situation and how they evaluate it (resulting in a certain state), farmers decide to take action. They can choose (a) to do business as usual and face the consequences, (b) to change to another pig quality level—provided that they have the required information—and (c) to quit pig farming.

In order to determine system behaviour, outputs of model versions are compared with respect to the total quality of pigs in the system as well as total farmer's satisfaction, determined by market conditions and farmer personality.

3 Methodology

For developing (versions of) the agent-based simulation model, we apply principles of incremental software design. For the specific agent-based development steps we follow state-of-the art ABM-modelling principles as laid down by Gilbert [9]. We designed a basic version of the agent-based model, the so-called base-ABM. This is a computer model containing all elements specified in the conceptual model. We then inspected the behaviour of the base-ABM by means of a sensitivity analysis that investigates all relevant parameters' and their combined values' effect on the observables. For this paper, we implement the selected what-if scenario in the base-ABM, resulting in a new model version (experimental ABM). We run the experimental ABM and compare the outputs with those from the base-ABM, with respect to the observables specified earlier. We interpret the results and draw conclusions for each of the research questions.

We implemented the base-ABM model according to the MASQ framework [8]. MASQ divides agent logic into Mind and Bodies, which communicate inside a Space (Fig.1). Culture gives the common patterns employed in minds for interpreting interactions that take place in a space. The main logic of an agent is located in its mind. The (one or more) bodies serve for communicating with other agents and (one or more) shared environments. Each body receives stimuli from a space, and produces actions in a space. Communication with other agents is mediated by the space. The idea in MASQ is that each different kind of interaction takes place in a specialized type of body. Thus one agent has one mind but as many bodies as necessary to define the different kinds of interactions.

For the model described here, we have three types of agents: pig farmers, SCN partners (buyers), and Livestock Bureau Officials (LBO). Currently, we have only elaborated the behaviour of the pig farmers. The other types of agents communicate with the pig farmers (mediated by spaces), but not with agents of their own type. Pig farmers communicate with other pig farmers in several ways for various purposes. For each kind of communication, a pig farmer agent has a separate body.

In our model implementation we use a message passing mechanism where all messages between agents are buffered in corresponding spaces. In this way the operations of all agents are decoupled and it is easy to organize the behavioural logic of agents. The effect of a body receiving a message is that the agent registers a change of state, which in turn influences all decisions made from that moment on. Decision making resides in the mind of the agent. So the mind interprets all information perceived through bodies and combines that information with its current state to alter the state and potentially send out information to spaces through one or more bodies.

Effectively, the possible actions of an agent are defined by the state of the agent when a message is processed—not necessarily when it is received. The full state of an agent is defined by the (values of) all its state variables. To specify behaviour, we distinguish two levels: at an aggregated level the kind of behaviour differs essentially between actions, whereas at a detailed level behaviour is expressed in terms of all state variables.

The model is set up as follows. Agent types are: farmer, buyer and LBO. Every month, each farmer has a fixed number of pigs of certain quality for sale. Quality is reflected as an integer value in the range [1, 100]. The quality is an indicator for the worth of the pig, i.e. its recommended price. Each month, every farmer tries to sell his pigs for this price. Whether a farmer succeeds depends on the demand: whether he can find a buyer who needs this quality. When demand is not sufficient, the farmer will not sell (all of) his pigs, and, as a result, he will be less satisfied. When his satisfaction drops below a certain threshold, he may decide to change his pig quality: either go up or down, dependent of demand. To go down in quality is without cost, but as a consequence, he will receive a lower income. To go up in quality requires that the farmer has enough know-how, expressed as units of information. With each information unit, the farmer's pigs quality level increases by 1. The farmer can obtain information units either from the LBO or from farmer friends.

Buyer agents try to fulfil a demand each month: buy an amount of pigs of certain quality. Demand is defined at system level. At system level, the model works with demand classes, e.g. low, medium and high, specified as contiguous subsets of the total quality range. For each quality class, a parameter at system level specifies the total number of pigs required in that class each month. During a model run, the total demand specified at system level is divided evenly over the buyers in the model. Buyers broadcast messages to all farmers, who may respond with an offer. The buyers evaluate the farmers' offers, and choose the best one, according to their criteria.

LBO agents reflect governmental influence. There is currently one LBO agent in the model. He visits a number of farmers each day, as specified by a parameter. The LBO carries a list containing all information units in the system. He supplies a number of information units to each farmer he visits, dependent on his support level (another parameter). Only new information adds up to a higher quality level. There are 100 unique information units in the system, equal to the maximum possible quality level. During the initialization of each model run, a number of information units is handed out to the farmers. The remaining units reside with the LBO.

Personality Characteristics of Agents

Agent types have personality characteristics that affect their evaluations and decisions. In contemporary psychology, the "Big Five" factors represent five broad domains or dimensions to describe human personality: Openness, Conscientiousness, Extroversion, Agreeableness and Neuroticism (OCEAN) [4]. Currently, our model only works with 'openness' (the tendency to try new things or to accept new ideas) and 'extraversion' (the tendency to share information with others). For farmer agents, personality affects their level of (dis)satisfaction, and consequently their decision to change quality. Personality also affects the probability that farmers will aim for a higher quality class (representing something unknown to them). For buyer and LBO agents actions are currently not affected by personality. LBO agents have a 'support level', which should be seen as a system level parameter, because it reflects a society's current practice for institutional support to farmers.

Cultural Aspects of Agents
For agent-based simulations, comparative models of culture that condense cultures into a limited number of basic issues, and assign comparative scores to cultures, are suitable modelling devices [13]. The most widely used and validated of these is the model by Hofstede [14], and currently consists of six dimensions: Identity, Hierarchy, Aggression and gender, Otherness and truth, Immutability versus pragmatism and Gratification of drives. The dimensions are social patterns, not personality traits.

At system level, our model supports investigating culture effects as well. The cultural dimensions parameters may affect evaluations, decisions and actions of agents. Currently, our model only employs the first dimension, Identity: individualism versus collectivism. This parameter affects the way farmer agents accept other farmers as friends, or select buyers to offer their pigs to: in a collectivistic population, farmers tend to prefer agents who belong to their group, whereas in an individualistic population farmers tend to prefer agents whose quality level is close to their own. Our model allows to divide agents over groups, the default being that they all belong to the same group.

4 Simulation Experiments and Results

One simulation run in our model reflects the passing of a number of months, specified by a parameter. Each month consists of a fixed number of ticks. Every month, agents choose to do specific actions. Each action costs time: choosing one means that there may not be enough time for another action. Buyers and LBOs have only one possible action, but farmers can choose between finding a buyer (sell pigs), improving quality (apply information), socializing (exchange information with farmer friends) and extending friends network (finding new friends). An action effectively leads to sending messages to other agents. After each tick, all messages are processed, resulting in possible state changes of the agents concerned.

Once every month, the farmers evaluate their situation. They update their satisfaction according to the reinforcement mechanism in 1. The value of E_t is based on how well they succeeded in selling their pigs, while their their personality determined the reinforcement weights (r^+ and r^-). If necessary, they may decide to aim for another quality class. To improve quality, they will have to check whether they have enough information units to go there. The effect of arriving at another quality class is that in the next month they will deal with different buyers.

$$\begin{aligned} S_t &= r^+ E_t + (1 - r^+) S_{t-1} \quad \text{if } E_t > 0 \\ S_t &= r^- E_t + (1 - r^-) S_{t-1} \quad \text{otherwise} \end{aligned} \tag{1}$$

As a base-ABM, we choose a scenario where demand is approximately 80% of supply, where lower quality is preferred over medium quality, and high quality is rarely demanded. We define quality classes as follows: low [1,40], medium [41,80] and high [81,100]. The base-ABM contains 10 farmers and 3 buyers (one in each quality class). The farmers all start in the lowest quality class. Every month they have 20 pigs for sale, so supply is 200 per month. The LBO visits 2 farmers each tick. The model runs for 30 months, 30 ticks per month. All other parameters are set to neutral default values.

Our experiments focus on demand variations, by changing the parameters for total demand in each class. Table 1 gives an overview of demand variations used; overall demand is the same for each demand class variation. Below the table, we give an interpretation of the findings.

Table 1 An overview of the demand variations, each time with 10 farmers, 3 buyers and 1 LBO. Total demand over all classes is equal for each experiment

Experiment	Total demand in Q-class 1	Total demand in Q-class 2	Total demand in Q-class 3
base	100	50	10
nr 1	0	0	160
nr 2	50	100	10
nr 3	30	60	70

- The base model experiment results in satisfied farmers, who are comfortable in Q-class 1 and do not move. See Fig. 2.
- In experiment 1, farmers do not move out of Q-class 1 either, but this time they are very dissatisfied. But it makes no sense to move to Q-class 2: there is no demand there. Q-class 3 is too far away from their starting situation.
- In experiment 2, the majority of the farmers moves to Q-class 2, because there is a high demand and it lies within reach. None of the farmers moves on to Q-class 3, mainly because they are satisfied in Q-class 2.
- In experiment 3, demand is going up along with Q-classes. The effect is that all farmers gradually increase their quality and change Q-class. The majority ends in Q-class 3. See Fig. 3.

Each day, the LBO visits a number of farmers. How many farmers the LBO visits per day is a system level parameter. We repeated the above experiments, with a value of 10 for this parameter, to reflect that the LBO can be fully informative to all farmers. The results of this change are especially interesting in the base situation and in experiment 3:

- In the base situation, a highly frequent LBO speeds up the process of moving up to the top quality level of class 1, but still no farmer advances to another Q-class. See Fig. 4.
- In experiment 3, a highly frequent LBO manages every single farmer to end up in Q-class 3. So the LBO effectively increased the number of farmers that moved to another Q-class. See Fig. 5.

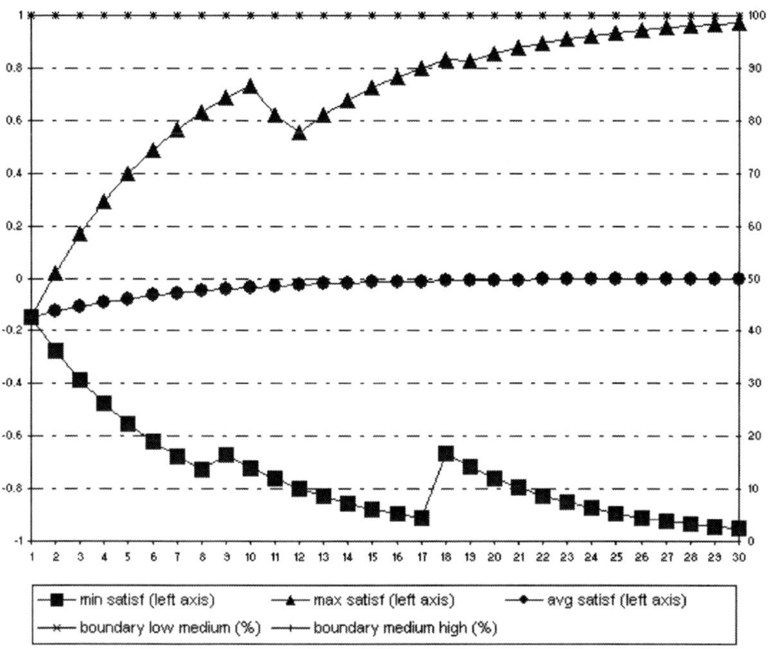

Fig. 2 Results of the base experiment: all farmers stay in Q-class 1 (low), so boundaries between quality classes are at 100% (small crosses at top of figure)

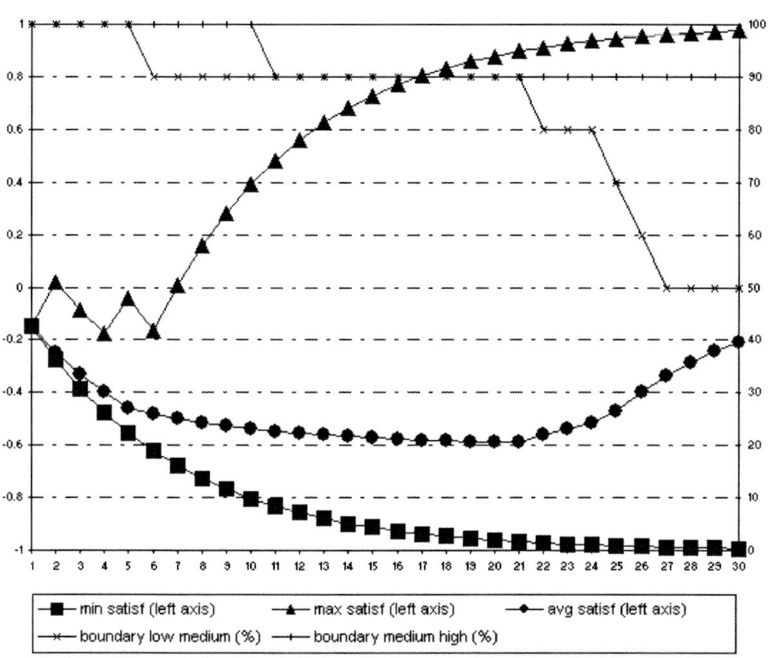

Fig. 3 Results of experiment 3: all farmers increase quality and some reach Q-class 3

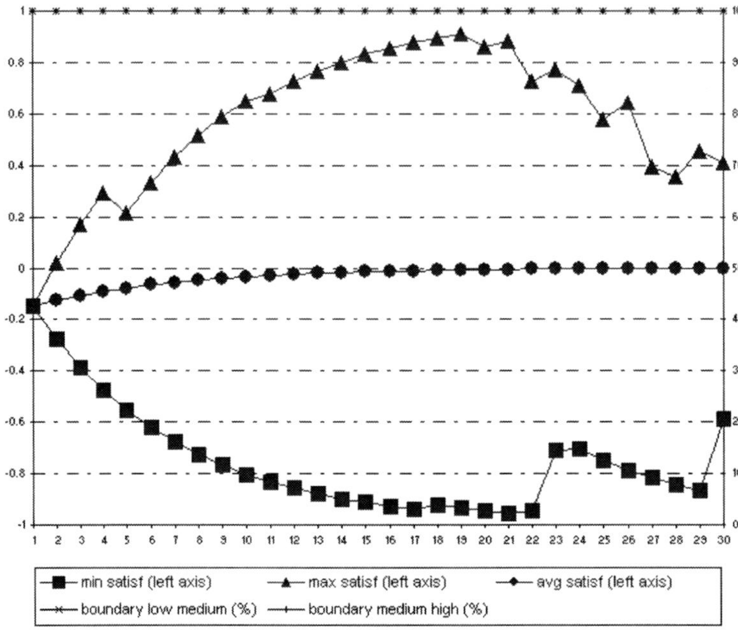

Fig. 4 Results of the base experiment but with visitation level 10

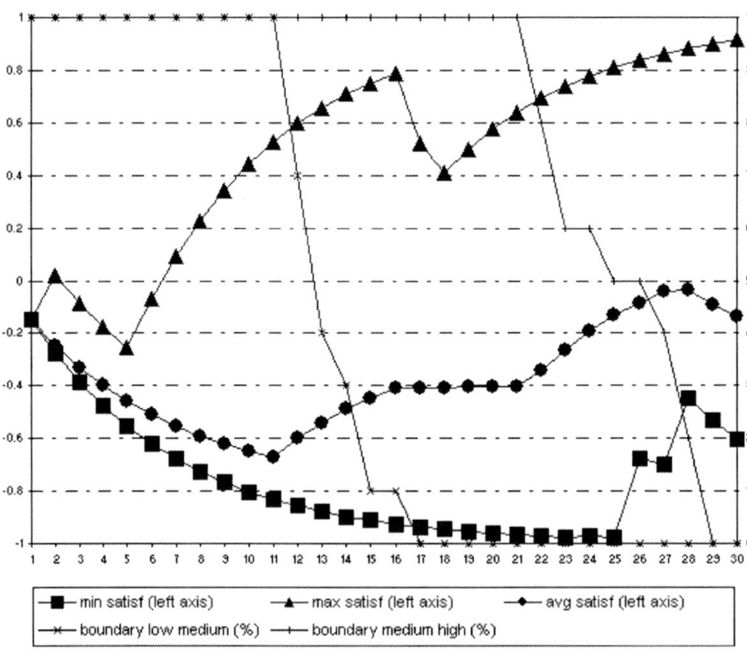

Fig. 5 Results of experiment 3 with visitation level 10: all farmers end up in Q-class 3 (high)

5 Conclusions and Discussion

This paper introduced agent-based modelling as a promising method to gain insight into the relationship between system level interventions and agent-level behaviour. We described simulation experiments that led to the insight that if a change of quality class is desired for a population of farmers, then the 'goal' demand should be posed in such a way that farmers have both the incentive and the possibility to move. A high demand in the goal class (higher than in the current class) serves as an incentive. The possibility to move implies that the goal class should be within reach of the farmers' current situation, since farmers can change quality only gradually.

The effect of increased LBO visitation level is high, especially in situations where demand already gives farmers an incentive to change quality class. In such cases, the LBO can make the difference for certain farmers who would stay behind without this extra 'know-how'. When there is no incentive in the demand situation, the LBO does not have so much influence. However, as the average quality within the quality class increases, the total quality in the system still increases.

Our research questions were: Can we adequately represent the real-world case, and: can we implement the selected scenario and how plausible are the effect(s) that we find in reality? The real-world case at hand was derived from a newspaper article referred to in the introduction of this paper, reporting large-scale dissatisfaction among the citizens. In our model, we could not include all factors leading to this event, but we could represent demand change in our experiments and we observed a low level of satisfaction (see experiment 1). Also the effect of exchanging information—one way or another—can be represented and investigated by means of our model.

In its current state, our agent-based model has rather rigid decision rules, and could benefit from substantial fine-tuning. Continuing work on our model will include further developing the agents' decision rules and actions repertoire, e.g. the earlier mentioned option to quit pig farming altogether, and elaborating on the effects of personality and culture. Strengthening the role of agent networks (farmer friends networks, but also buyer-farmer and buyer-buyer networks) will presumably add interesting dynamics to the agents population and simulation outcomes.

References

1. Brown, J.S., Duguid, P. (2002) The social life of information. Harvard Business School Press, Boston
2. China: White paper (2007) The quality and safety of food in China. State Council Information Office. URL `http://www.chinadaily.com.cn/bizchina/2007-08/18/content_6032837.htm`
3. ChinaDaily: Pork retailers in 6th day of strike (2010). URL `http://www.chinadaily.com.cn/bizchina/2010-01/07/content_9277866.htm`
4. Costa P. T., J., McCrae, R.R. (1992) Revised NEO Personality Inventory (NEO-PI-R) and the Five Factor Inventory (NEO-FFI): Professional Manual. Psychological Assessment Resources Inc., Odessa, Florida

5. Dignum, V. (2004) A model for organizational interaction: based on agents, founded in logic. Ph.D. thesis, Utrecht University
6. Dimitriadis, N.I., Koh, S.C.L. (2005) Information flow and supply chain management in local production networks: the role of people and information systems. Production Planning & Control **16**(6), 545–554
7. Fabiosa, J.F., Hu, D., Fang, C.: A case study of China's commercial pork value chain. Tech. rep., Midwest Agribusiness Trade Research and Information Center (MATRIC), Iowa State University (2005)
8. Ferber, J., Stratulat, T., Tranier, J. (2009) Towards an integral approach of organizations: the masq approach. In: V. Dignum (ed.) Multi-agent Systems: Semantics and Dynamics of Organizational Models. IGI
9. Gilbert, N. (2008) Agent-Based Models, *Quantitative Applications in the Social Sciences*, vol. 153. Sage Publications, Los Angeles
10. Hamill, L., Gilbert, N. (2009) Social circles: A simple structure for agent-based social network models. Journal of Artificial Societies and Social Simulation **12**((2)3)
11. Harlow, A.A. (1960) The hog cycle and the cobweb theorem. Journal of Farm Economics **42**(4), 842–853
12. Hofstede, G., Hofstede, G.J. (2005) Cultures and Organizations: Software of the Mind. Third Millennium Edition. McGraw-Hill, New York
13. Hofstede, G., Hofstede, G.J., Minkov, M. (2010) Cultures and Organizations: Software of the Mind. McGraw-Hill, New York
14. Hofstede, G.J., Jonker, C.M., Verwaart, D. (2010)Computational modelling of culture's consequences. In: MABS 2010; 11th international workshop on multi-agent based simulation, Toronto
15. Isaac, M.E., Erickson, B.H., Quashie-Sam, S.J., Timmer, V.R. (2007) Transfer of knowledge on agroforestry management practices: the structure of farmer advice networks. Ecol. Soc. **12**(2)
16. Jahiel, A.R. (1998) The organization of environmental protection in china. China Q. **156**, 757–787
17. Lu, H. (2007) The role of guanxi in buyer-seller relationships in China: A survey of vegetable supply chains in Jiangsu Province. International Chains and networks series. Wageningen Academic Publishers
18. Moss, S. (2002) Policy analysis from first principles. Proc. Natl. Acad. Sci. U. S. A. **99**, 7267–7274
19. Narrod, C., Delgado, C., Tiongco, M. (2006) Socio-Economic Implications of the Livestock Industrialization Process: How will Smallholders Fare? Topic under Livestock Industrialization Process on Smallholders- (WC 11,777), a book chapter on Livestock in a Changing Landscape: Drivers, Consequences, and Responses. FAO, Rome
20. Osinga, S.A., Roozmand, O., Kramer, M.R., Hofstede, G.J. (2010) An agent-based model of information management in the chinese pig sector: top-down versus bottom-up. In: Wageningen International Conference on Chain and Network Management. Wageningen Academic Publishers. To appear

Part VI
Macroeconomics

Wealth Distribution Evolution in an Agent-Based Computational Economy

Victor Romanov, Dmitry Yakovlev, and Anna Lelchuk

Abstract In this paper we study the modification of wealth distribution among the customers during quite a long period of time in the model — several model years. During this time customers get their income in forms of salary depending on enterprise production volume and assortment, or redundancy payments. As a part of the study it was detected that whilst the initial wealth distribution was uniform a strong non-uniformity arises after several years in the model.

The model includes the following interacting agent classes: customer, bank, labor market, state, enterprise, market, university, and mass media. The model also allows us to evaluate the relations among the efficiency of enterprises' investment strategies, tax level and customer's prosperity and unemployment level. The possibility of obtaining a new specialty by a fired agent for the purpose of stabilization and increasing his profit and improve standard of life is considered in the paper as well.

1 Introduction

The problems of multi-agent modeling in economics have been attracting more and more scientific attention, since the Leigh Tesfatsion pioneer works, accumulated in [18]. Multi-agent simulation for food market is considered in the paper [5]. The

Victor Romanov
The Plekhanov Russian Academy of Economics, Moscow, Russia
e-mail: victorromanov1@gmail.com

Dmitry Yakovlev
The Plekhanov Russian Academy of Economics, Moscow, Russia
e-mail: jdemon@yandex.ru

Anna Lelchuk
The Plekhanov Russian Academy of Economics, Moscow, Russia
e-mail: lelc-anya@yandex.ru

M. LiCalzi et al. (eds.), *Progress in Artificial Economics*, Lecture Notes in Economics and Mathematical Systems 645, DOI 10.1007/978-3-642-13947-5_16,

related topics concerning negotiations are discussed in the article [13]. In this paper we are particularly interested in enterprise simulation and consumer behavior issues partly covered in the articles [11].

The problem of inequality of wealth distribution in the society has been considered in scientific literature for quite a long time. Wealth distribution in artificial and real economics has attracted great attention in the research community since the last decades of the XIX century, starting with Vilfredo Pareto [12]. A brilliant analytical review on this topic is contained in the articles [10, 8].

The goal of our paper is to study how the customers' wealth distribution function is modified during their lifetime. The wealth is formed by incoming salary or redundancy payments and depends on the variability of demand and on the corresponding modified enterprise production volume and assortment. By customers' wealth we shall basically mean the cumulative sum of customers' money composed of the initial wealth endowed in a random manner according to the predefined law, and the amount of the following regular earnings (salary or redundancy relieves) minus cost of the purchases made.

To make sure that the results are robust, in this research we adduce the results of a wealth distribution simulation in the virtual economics involving from 300 to 1000 customers, and 5 and 10 enterprises respectively. We discovered the change of distribution function according to the scheme: rectangular — normal — gamma — normal — rectangular during a twelve-year period. We also found out that a group of customers with a much higher income (5–6 times higher than an average income), and a gap dividing the two groups of customers appear once we set some critical tax level (24%).

2 Equilibrium Wealth Distribution Models

One of the earliest works devoted to the problem of inequality based on multi-agent approach was that of Angle [3]. The evolution of wealth distribution between two types of agents — winner and looser — takes place in the model of Angle.

The model demonstrated that the process leads to a gamma distribution.

Different kinds of stable probability distribution of agents concentration were discovered in the ants model by Kirman and described by Alfarano *et al.* (2005), (2008) [1, 2]. In the model the ant colony is inhabited by N agents, each in one of two possible states. The average number of neighbours per agent is D, relative communication rate is D/N. Transition rate from one state to another depends on the concentration n/N and $(N-n)/N$ of agents in the appropriate state with strength random state transition strength coefficient a, and on presence of interaction among agents $n(N-n)$ with strength coefficient b. Markov chain approach based on the evolution equation for transition $w(x,t/x_0,t_0)$ probability density was applied.

The state of the system depends on the concentration of agents in either of the two states. Equilibrium state refers to stationary distribution of the process according to the time the system spends in the state x.

According to the general solution well known from handbook [17], we get

$$w_{st}(x) = \frac{\Gamma(2\varepsilon)}{2^{2\varepsilon-1}\Gamma(\varepsilon)^2}(1-x^2)^{\varepsilon-1}, \varepsilon = \frac{a}{b}.$$

The remarkable characteristics of this expression is that for $\varepsilon > 1$ the function is unimodal, for $\varepsilon < 1$ it is bimodal, for $\varepsilon = 1$ it is uniform, and for $\varepsilon >> 1$ the equilibrium distribution converges to Gaussian.

The approach from the point of view of classical statistical mechanics and thermodynamics was applied by Victor Yakovenko *et al.* [7, 4], who offered the model that included agents (particles) exchanging a constant amount of energy (money). They studied probability distribution of money and income for ensemble of economic agents. The result of income distribution interpolated between exponential Boltzmann-Gibbs law for average and low income, and power law for high income.

In the trading model [6] propensity to save is assigned to all agents; they save λ-fraction $(0 \leq \lambda < 1)$ of wealth during the trade. Wealth distribution in this model converges to the gamma distribution.

Silver J. *et al.* [16] describes a market consisting of many agents with Cobb-Douglas preferences in the case of two goods. They describe mathematically that the gamma distribution arises for a broader class of preference distributions in the limit of large numbers of individual agents.

A brief review of researches on the exchange process modeling in multi-agent systems demonstrates that in the different models the same set of available probability distribution functions arises: exponential, Gaussian, gamma, Pareto. This probability distributions may turn from one form into another with different values of model parameters. The results obtained in our work do not contradict the research made by other authors. The specific feature of our model is that such transition happens during one cycle of model functioning.

The wealth distribution evolution in our model may be explained by changing of drift and diffusion coefficients in time.

The plot of the finite difference approximation time dependent drift $\frac{d<x>}{dt}$ and diffusion $\frac{d<x^2>}{dt}$ coefficients are presented in Figs. 1 and 2, respectively.

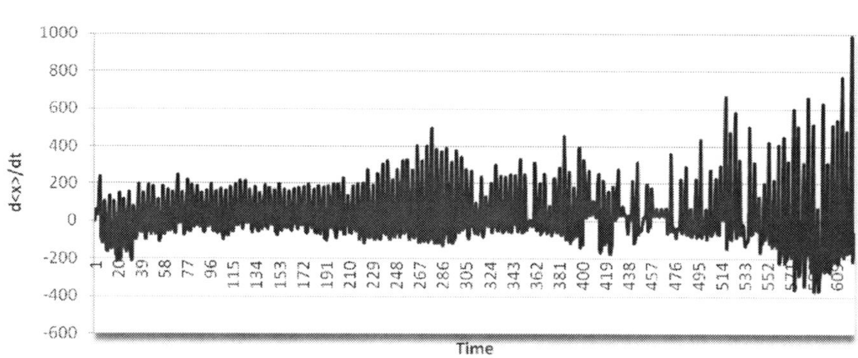

Fig. 1 Drift coefficient

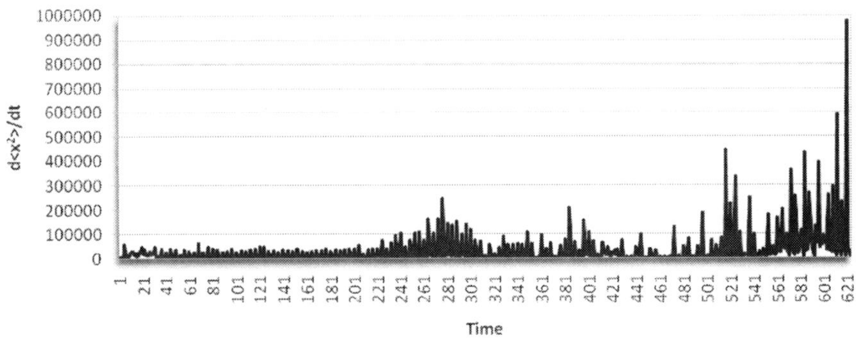

Fig. 2 Diffusion coefficient

We suppose that in our model transition from one to another kind of probability distribution function in the cycle of model operation happens due to non-linear type of dependences and positive feedback.

To illustrate this proposition let us consider the following non-linear Fokker-Planck equation [19]:

$$\frac{\partial w(x,t)}{\partial t} = \frac{\partial(\alpha x + \beta\langle x(t)\rangle w(x,t))}{\partial x} + D\frac{\partial^2 w(x,t)}{\partial x^2},$$

where $\langle x(t)\rangle = \int\limits_{-\infty}^{\infty} xw(x,t)dx$, and α, β, and D are constant.

Assuming initial δ-function wealth distribution $w(x,0) = \delta(x - x_0)$ stable probability distribution "moves" according to $\langle x(t)\rangle$ changing in time and gets modified due to dependence of σ on t. So in case of non-linear Fokker-Planck equations we may expect evolution of probability distribution in time.

3 Model Architecture

The key issues in designing an adequate and effective model of economy are to understand and analyze the following interrelated phenomena:

- wealth possessed by customers and its repartition during their lifetime;
- effect of demand on production level;
- dependence of model's macroeconomic indices on tax level and the enterprise strategy;
- impact of the state's tax policy on macroeconomic indices;
- relationship between enterprise investment strategy and such macroeconomic indicators as tax level, employed population as well as university fee;
- dependence of the amount of working population on the enterprise's investment strategy, tax level and university fee.

While doing this research the authors presumed that the changing composite demand of the economic agents is the major driving force of the economic development. We also believe that economic efficiency is mainly determined by the correlation of the current output with the varying consumer needs and demand.

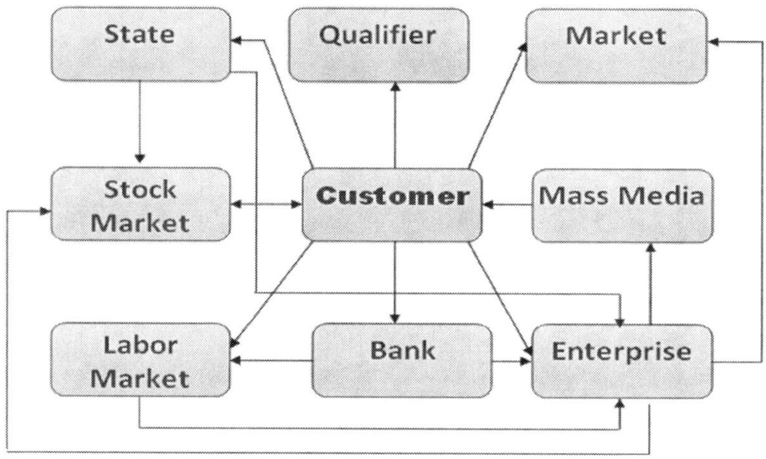

Fig. 3 Classes diagram

Classes diagram of the model actors is presented in Fig. 3. Initially the diagram was created in Altova UModel but here we present a simplified scheme of model's classes.

- ENTERPRISE acts as a seller on the market, a client of the bank, a taxpayer and an employer, makes also an initial public stock offering;
- MARKET provides customers with an access to goods produced by enterprises and in such a way keeps the dynamic balance between the aggregate demand and aggregate supply, and forms an equilibrium price on the base of bargaining;
- STATE collects taxes from the enterprises and pays redundancy relieves;
- BANK (FIRST_NAT_BANK) organizes money transactions among agents;
- LABOR_MARKET lists vacancies that enterprises possess and sends messages about them to the job seekers, and provides equilibrium between labor demand and supply;
- QUALIFIER is an educational institution that performs an employee's training and requalification for a definite fee;
- MASS MEDIA receives orders from the enterprises for an advertising campaign, then selects a random sample of customers with the aim to increase the value of the subjective utility function of the advertised product;
- STOCK MARKET provides the buy/sell operations among the issuers and the shareholders.

The model interface is presented in Fig. 4.

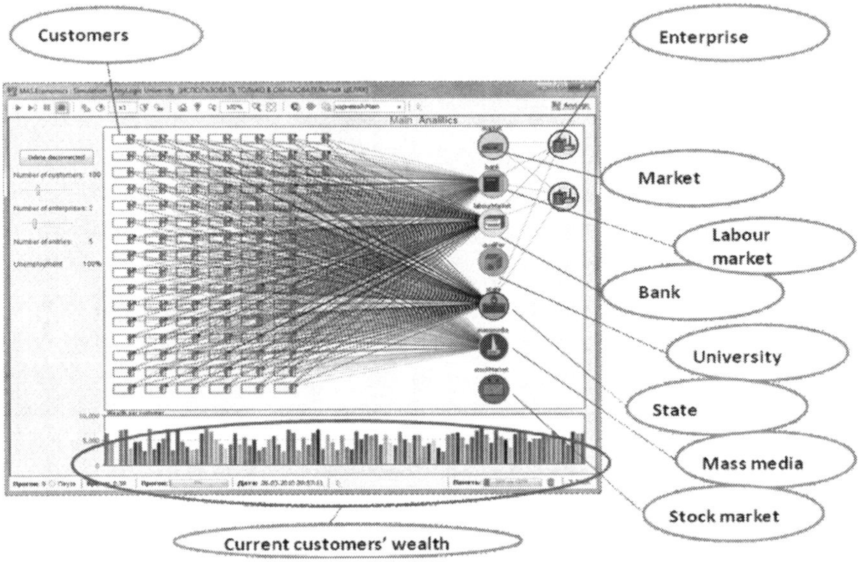

Fig. 4 The model interface

The process of model's dynamics is determined by three main types of agents' activities:

- Agent job placement;
- Customer behavior;
- Enterprise investment strategy;
- Current wealth distribution among customers.

4 Agent Job Placement

A customer may be in one of four possible states: employed, unemployed, student, dead. At the initial stage all the customers are unemployed and get redundancy payments from the state. All of them send messages to the labor market and are registered there as unemployed. At the same time the enterprises send messages to the labor market with the vacancies they possess. Next, the agents get employed on one of the enterprises and stop receiving the redundancy relief if the number of vacancies is equal to the number of unemployed; otherwise a customer may remain unemployed. As employees they obtain salary according to their degree and qualification.

In the case of reduced production level an enterprise can fire an employee so he registers at the labor market again.

5 Customer Behavior

Customer h, $h = 1,2,\dots,H$ may be described by initial capital C_h and consuming profile F_{hi}. Customer may be also characterised by qualification and degree. Qualification L_i is defined by the kind of production he is involved in. Assume that there are three levels of degree: degree1, degree2, degree3. Consumption profile has the following restrictions: consumed product cost must be less than or equal to $k_1 \cdot B_h$, where B_h is the income which includes salary, dividends and profits from stock market operations or redundancy relieves (for those who are unemployed).

The volume of the consumed products combination should not be lower than customer's survival level. A customer adjusts his level of consumption to his profits. Different products/consumption level is defined by functions such as sine wave (seasonal fluctuation), large period sine wave (age fluctuation), almost constant function, exponential decreasing function, exponential increasing (popular) function, logical function.

Current wealth is distributed in the following way:

- Taxes make up γB_h, where γ — taxes rate, $0 \le \gamma \le 1$;
- Current expenses βB_h , where β — $0 \le \beta \le 1$;
- Savings, aimed for requalification or education. Amount, remaining after taxes paying $(1-\gamma)B_h$. For current expenses $\beta(1-\gamma)B_h$. $C_{edu} = 1-\beta(1-\gamma)B_h$ remains for education. Market basket cost of the customer h for period T is $\sigma_h = \sum_{t\ 1}^{T}\sum_{i\ 1}^{N}\phi_{it}F_{hit}$ and should not exceed βB_h, i.e. $\sigma_h \le \beta B_h$. If this condition is not satisfied the customer adjust his consumption the following way: $F_{hit+1} = F_{hit} + \alpha(\sigma_h - \beta B_h)$, and checking survival condition $E_h = \sum_{t-1}^{T}\sum_{i-1}^{N} F_{hit} \cdot C_i \ge E^0$.

If this condition is not satisfied during the period T_{max} the customer leaves the model. In the case when the customer is unemployed he can browse the vacancies list on the labor market. If he finds a vacancy of the enterprise j corresponding to his qualification in the list, he can make a contract with this enterprise, and after that he will be included into the salary list. If he buys enterprise's shares he will also be included into the list of those who obtain dividends. To improve his level of earning he may increase his degree (if it is not maximum) using qualifier's service and paying the necessary sum S_{edu} under condition that $C_{edu} \ge S_{edu}$. If there is no vacancy for his qualification in the vacancies list he can get a new qualification using qualifier's services.

The aim of each customer is to increase his quality of life. The quality of life in the model consists of wealth and value of utility function corresponding to a customer's individual consumption profile. To increase his quality of life a customer uses his beliefs. According to them he can improve his quality of life by the following actions:

- constant job seeking according to a customer's qualification and degree (including requalification);

- buying/selling shares on the stock market and getting the dividends;
- saving money in a bank account;
- becoming an enterprise owner by buying its controlling interest

The customer strategy is the following. First of all the customer makes up a list of products that he can afford to buy at a definite sum not depending on whether he needs it or not. After that he sorts the price list in descending order by the value of utility function. Then, using this sorted list, the customer buys a product that gives the maximum increment of utility function and acts like that until he depletes the available financial resources.

6 Stock Market Functioning

Stock market simulation was previously described in the papers [14, 15] which, in turn, were based on the model described by Li and Rosser [9]. New factors were introduced in the mentioned model. "Bad" or "good" news arises every moment. There exists the memory in the model, which determines the rate at which news can be forgotten. In such a way, the news background is being formed by addition of decaying news intensities. The news background modifies the fundamental value of current market price according to the logistic law. The insiders at moment $t-1$ know the prices at the moment t. The market simulation by means of the proposed model shows that, in case of "good" news, the stock-market prices are rising, and in case of "bad" news, the prices are falling. Moreover, the parameter that determines the news-forgetting rate changes the picture of rising and falling prices. The model also shows, that the effect of insiders' activity depends on the return volume extracted by him, and when insiders' return approaches some crucial value, the fundamental value v abruptly falls down, and further with the increasing insiders' pressure the market explodes.

Stock market involves purchase and sale (trading) of securities emitted by enterprises. Each enterprise can emit only one kind of security. As different types of securities are traded independently on the stock market we can describe the trading process as if there was only one type of securities on the market. According to his trading strategy a trader can belong to one of the three categories — fundamentalists, chartists (noise traders), and insiders (in special cases). Fundamentalist traders may change their strategy and pass into chartist category and vice versa in case the former strategy does not bring them enough profit.

To capture the significant contribution of news analysis to the decision making, we added the news background to our model, that replaced or was added to the "market noise system". For each day, k random news events ξ_k are generated, whose value (or intensity) depends on the position of the "good-bad" slider, and the news sign ("bad" or "good") depends on the slider displacement from the neutral position. The news in this model has the property of being accumulated in time, but the strength of news is decaying with time. The accumulated news comprise the news background, which may be neutral, positive, or negative, depending on the sign of accumulated news.

The term "insiders" reflects the fact that the specific information they possessed until the current moment was for internal use only inside a limited number of users. As a result, these persons, getting the insiders' information before its official publication, have an opportunity to make the certain moves at the stock market for obtaining superior profits.

7 Enterprise Investment Strategy

It is considered that the agent ENTERPRISE acts in perfect competition.

All the enterprises in the model produce the same types of goods at each interval. Some of them belong to the group of basic consumer goods (food, clothes), others are articles of secondary necessity (mobile phones) and others are luxury goods (automobiles, cottages). In the model we use the Cobb-Douglas production function. An enterprise is trying to find an optimal volume of output taking into account the balance between total income and total costs. An enterprise accomplishes its investment strategies in an effort to remove the mismatch between demand and supply on the market.

The customer driven enterprise strategy in the model is realized as follows. Three activity partitions display three types of enterprise strategies. The first one corresponds to the case when all the profit obtained by an enterprise is divided among the employees according to their qualification and degree. The second one shows the case when an enterprise gets a positive profit and uses it to increase the production level according to the customers' demand. The third one describes the situation when an enterprise has a negative profit and it covers its losses by partly selling its capacity. If nevertheless the losses are not covered and the enterprise's capital becomes negative the enterprise is declared bankrupt and leaves the model.

8 Results

The model is implemented and simulated within the AnyLogic environment [20].

To present the results of the research on the modification of the wealth distribution function we examined condition of wealth distribution during the first twelve years of model run with the following initial parameters: 300 customers and 5 enterprises in the model. The data on the customers' wealth were imported and processed in Statistica 8.0 (http://statsoft.com/). The following figures show how the wealth distribution changes from the initial period to the twelfth year (624 model time steps).

After 52 time steps that makes one year the wealth distribution modifies from rectangular to a normal one as you can see in Figs. 5 and 6:

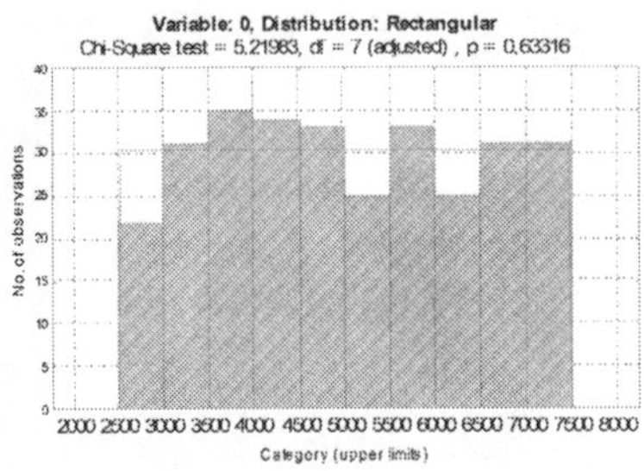

Fig. 5 Original wealth distribution

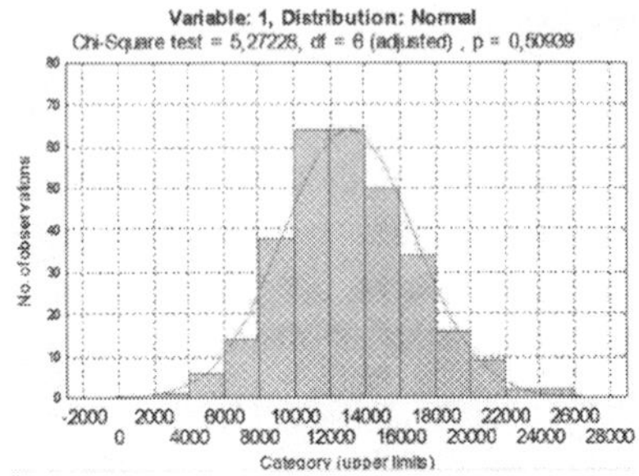

Fig. 6 Wealth distribution one year later

The third year data does not appear to belong to any particular distribution type (Fig. 7).

During another nine model years we'll follow up how that picture changes. The very middle of the time period analyzed in the model, six years, is characterized by a gamma wealth distribution. It is evident from Fig. 8 that wealth is mainly accumulated by the middle class customers whilst the number of people belonging to the poor class declines and the wealthy class, on the contrary, increases. This trend is preserved for the several years to come.

As you can see in the Fig. 9 (that shows the condition of the model in the eight year) the wealth distribution switches back to normal as it was recorded seven years before that. Again, the middle class as well as the wealthy class are growing unlike the poor one.

What we can observe in the last chart (Fig. 10) is that the wealth distribution returns to the uniform distribution after twelve years of running the model.

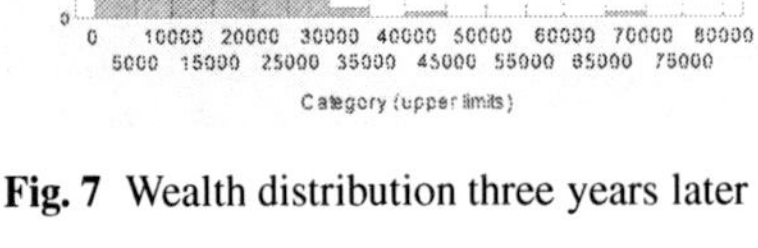

Fig. 7 Wealth distribution three years later

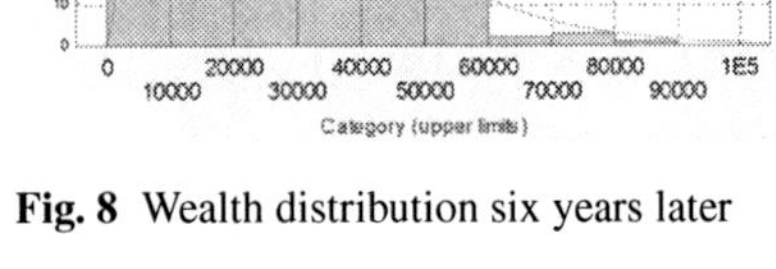

Fig. 8 Wealth distribution six years later

So we can observe the following steps in the evolution of the wealth distribution: uniform — normal — gamma — normal — uniform. As the production development and assortment innovations are not introduced in the model the system becomes stabilized as well as the wealth distribution.

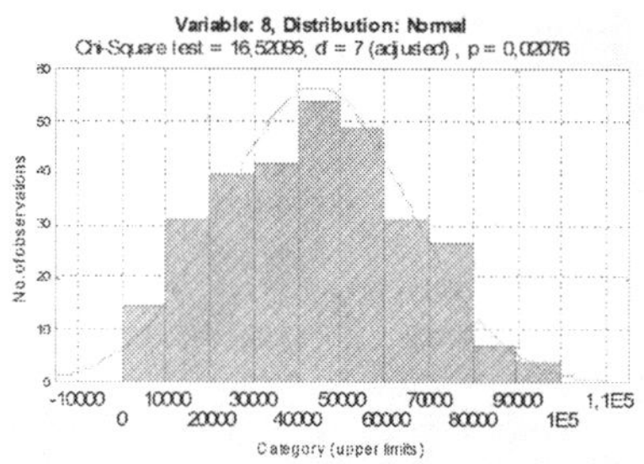

Fig. 9 Wealth distribution eight years later

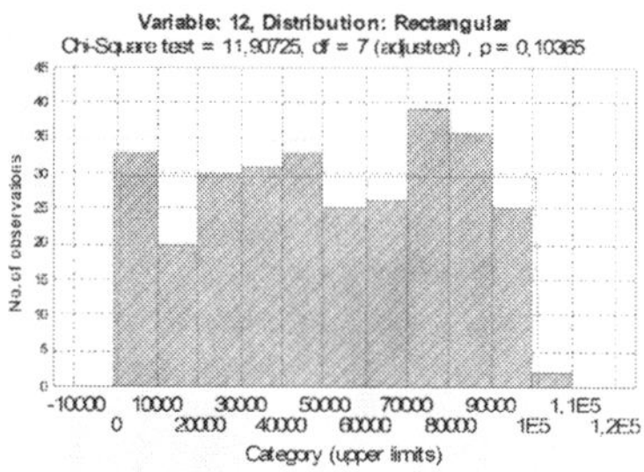

Fig. 10 Wealth distribution twelve years later

We also found out that a group of customers with a much higher income (5–6 times higher than an average income) and a gap dividing the two groups of customers appears once we set some critical tax level (24%). This result is proved out both if we set 300 customers (5 enterprises), 600 customers (10 enterprises), and 1000 customers (10 enterprises) initially. The effect can be better observed with the initial parameters of 600 customers and 10 enterprises after twelve model years.

A similar influence of tax level on the type of distribution function is mentioned in the work of Chakrabarty [6]. In a trading model with taxation and redistribution where τ is the fraction of money taxed from traders it was shown [14] that the exponential distribution transfers to the gamma function as τ goes up and then after a threshold, it returns to the exponential for higher values of τ.

References

1. Alfarano, S., Lux, T., Wagner, F. (2005) Time-Variation of Higher Moments in a Financial Market with Heterogeneous Agents: An Analytical Approach. Journal of Economic Dynamics and Control 32(1): 101-136
2. Alfarano, S., Milakovic, M., Raddant, M. (2008), Network Hierarchy in Kirman's Ant Model: Fund Investments Can Create Risks. `http://www.bwl.uni-kiel.de/vwlinstitute/gwif/files/papers/ncore.pdf`
3. Angle, J. (1986) The surplus theory of social stratification and the size distribution of personal wealth. Social Forces 65(2): 293
4. Banerjee, A., Yakovenko, V.M. (2009) Universal patterns of inequality. `http://econpapers.repec.org/scripts/search/search.asp?ss=3&adv=t&sslg=AND&ni=&nit=epdate&mh=100&sort=rank`
5. Caillou, P., Baptista, T., Curchod, C. (2008) Multi-agent Based simulation for Decision-Making: an application to Rungis food market. In: Climaco J, Kersten G, Costa J P (eds.), GDN 2008, Coimbra, Portugal, 193-194
6. Chakraborti, A., Chakrabarti, B. (2000) Statistical mechanics of money: How saving propensities affects its distribution. European Physical Journal B 17:167-170
7. Dragulescu, A.A., Yakovenko, V.M. (2000) Statistical mechanics of money, income, and wealth. European Physical Journal B 17:723-729

8. LeBaron, B. (2002) Building the Santa Fe Artificial Stock Market. Working Paper, Brandeis University, `http://www.econ.iastate.edu/tesfatsi/blake.sfisum.pdf`
9. Li, H., Rosser, B.J., Jr, (2001) Emergent volatility in asset markets with heterogeneous agents. Discrete Dynamics in Nature and Society 6 (3):171-180
10. Lux, T. (2008) Application of Statistical Physics and Finance and Economics, University of Kiel, Working paper 1425, `http://www.ifw-members.ifw-kiel.de/publications/applications-of-statistical-physics-in-finance-and-economics/KWP_1425_Applications%20of%20Statistical%20Physics.pdf`
11. Nimis, J., Stockheim, T. (2004) The Agent. Enterprise Multi-Multi-agent System. In: Bichler, M., Holtmann, C., Kirn, St., Weinhardt, C. (eds.) Coordination and Agent Technology in Value Networks. Proceedings of the Conference on Agent Technology in Business Applications (ATeBA 2004), part of the Multi-Conference on Business Information Systems (MKWI 2004). Essen, Germany
12. Pareto, V. (1897) Cours d'Economie politique, F. Rouge, Lausanne
13. Praca, I., Viamote, M.J., Ramos, C., Vale, Z., (2008) A Multi-Agent Market Simulator to Support Negotiation Decision Making, `http://www.gecad.isep.ipp.pt/report0307/Papers/IESPublications/eChallenges.pdf`
14. Romanov, V., Naletova, O., Pantileeva, E. (2006) The simulation of news and insiders' influence on stock-market prices dynamics in non-linear model. Computational Finance and its applications II, WIT Press, 309-328
15. Romanov, V., Slepov, V., Badrina, M., Federyakov, A. (2008) Multifractal analysis and multi-agent simulation for market crash prediction. Computational Finance and its applications III. WIT Press, 13-22
16. Silver, J., Slud, E., Takamoto, K. (2001) Statistical Equilibrium Wealth Distributions in an Exchange Economy with Stochastic Preferences. Journal of Economic Theory 106(2):417-435
17. Stratonovich, R. L. (2009) Stochastic processes in dynamic systems (in Russian), Moscow
18. Tesfatsion, L., Judd, K. (2006) Handbook of Computational Economics, vol. 2, North-Holland
19. Zhang, D.S., Wei, G.W., Kouri, D.J., Hoffman, D.K. (1997) Numerical method for the non-linear Fokker-Planck equation. Physical review E 56(1): 1197-1206
20. `http://www.xjtek.com/`

Endogenous Credit Dynamics as Source of Business Cycles in the EURACE Model

Andrea Teglio, Marco Raberto, and Silvano Cincotti

Abstract The paper investigates the relationship between the amount of credit money in the economy and the variability of output and prices in the EURACE model. First we examine if the decision about dividends payment by the firms can affect this variability, then we adopt the policy measure of quantitative easing, that has been largely used by the Fed and the Bank of England during the recent crisis, in order to understand its effect on economic instability. Results show the emergence of endogenous business cycles which are mainly due to the interplay between the real economic activity and its financing through the credit market. In particular, the amplitude of the business cycles strongly raises when the fraction of earnings paid out by firms as dividends is higher, that is when firms are more constrained to borrow credit money to fund their activity.

1 Introduction

The aim of this paper is to analyze the reasons of output and price variabilities in the EURACE economic environment. The issue is of primary importance, given that reducing macroeconomic volatility has numerous benefits. Lower volatility of inflation improves market functioning, makes economic planning easier, and reduces the resources devoted to hedging inflation risks. Lower volatility of output tends to imply more stable employment and a reduction in the extent of economic uncertainty confronting households and firms.

Agent-based computational economics (ACE) has been characterized by a great deal of development in recent years (see [21] for a recent survey). There have been many studies regarding finance (see [17] for a review), while others have focused on

Andrea Teglio
Departament d'Economia, Universitat Jaume I, Av. Sos Baynat, Castellón de la Plana, Spain
e-mail: teglio@eco.uji.es

Marco Raberto
School of Science and Engineering, Reykjavik University, Menntavegur 1, 101 Reykjavik, Iceland
e-mail: raberto@ru.is

Silvano Cincotti
DIBE-CINEF, Università di Genova, Via Opera Pia 11a, 16145 Genova, Italy
e-mail: cincotti@dibe.unige.it

M. LiCalzi et al. (eds.), *Progress in Artificial Economics*, Lecture Notes in Economics and Mathematical Systems 645, DOI 10.1007/978-3-642-13947-5_17,

labour and goods market [18, 20] or industrial organization [9]. However, only a few partial attempts have been made to model a multiple-market economy as a whole [3, 1]. In this respect, the EURACE simulator is certainly more complete, incorporating many crucial connections between the real economy and the credit and financial markets. In order to understand the recent crisis, and in general to understand the profound functioning of modern economies, we think it is not possible to ignore that connections any more.

The aim of our paper is twofold. First we investigate if the decision about dividends payment by the firms can affect the variability of output and prices, and our results are clearly affirmative in this respect. Then we adopt the policy measure of quantitative easing, that has been largely used by the FED and the Bank of England during the recent crisis, in order to understand its effect on economic instability.

In concrete terms, our experiments on the EURACE platform consist of different simulations for different parameter values. We take into consideration the effects of two critical parameters of the model.

The first one, as said above, regards the financial management decision making of the firms, and corresponds to the fraction of net earnings paid by the firm to shareholders in form of dividends. The dividends decision impacts on many sectors of the model. In the financial market, for instance, agents beliefs on asset returns take into account corporate equity and expected cash flows, establishing an endogenous integration between the financial side and the real side of the economy. In particular, fundamentalist trading behavior is based on the difference between stock market capitalization and the book value of equity, therefore generating an interaction between the equity of the firm and the price of its asset in the financial market. Concerning the credit market, the dividends payment is strongly correlated with the loans request of the firm and consequently influences the amount of credit created by the commercial banks; as our results show in Sect. 3, the credit amount proves to be decisive for its effects on the variability of output and prices.

The second parameter of our study is a binary flag that activates the possibility for the central bank to buy treasure bills in the financial market, when a government asks for it. In practical terms, the central bank expands its balance sheet by purchasing government bonds. This form of monetary policy, widely adopted during the global financial and economic crisis of the years 2007-09, which is used to stimulate an economy where interest rates are close to zero, is called quantitative easing (QE). The creation of this new money is intended to seed the increase in the overall money supply through deposit multiplication by encouraging lending by these institutions and reducing the cost of borrowing, thereby stimulating the economy. Besides, quantitative easing is intended to help the funding of government budget deficit, by reducing the cost of debt as well as reducing the risk of debt rolling over.

The paper is organized as follows. In Sect. 2 it is given an overall description of the model with particular attention to the features that are relevant to this article. Section 3 presents the computational results of our study and a related discussion.

2 The Model

The EURACE model represents a fully integrated macroeconomy consisting of three economic spheres: the real sphere (consumption goods, investment goods, and labour market), the financial sphere (credit and financial markets), and the public sector (Government, Central Bank and Eurostat).

Given the complexity of the underlying technological framework and given the considerable extension of the EURACE model, it is not possible to present within this paper an exhaustive explanation of the economic modelling choices, together with a related mathematical or algorithmic description. Consequently, we will limit our approach to a general qualitative explanation of the main key features of the model, treating in a concise way each different market, and giving prominence to those modelling aspects that attain to the argument of the specific analysis we are presenting in this paper.

If the reader needs more details about the EURACE implementation, he will find a quite exhaustive summary in [10]. Moreover, when needed, we will cite specific EURACE deliverables. Some general information on EURACE can be found in [8].

Both the modelling of agents behaviors and the modelling of markets protocols are empirically inspired by the real world.

Agent decision processes follow the usual and realistic assumptions of agent-based economics about bounded rationality, limited information gathering and storage capacities, and limited computational capabilities of the economic agents; see e.g. [21] for a recent survey on this approach. These assumptions lead us to use simple heuristics to model the agents' behaviour, derived from the management literature for firms, and from experimental and behavioural economics for consumers/investors [7, 2]. We also make use of experimental evidence from the psychological literature on decision making. For example, the modelling of households' portfolio decisions on the financial market is based on Prospect Theory (see [22]).

The rules used by the agents are simple but not necessarily fixed. Their parameters can be subject to learning, and thus adapted to a changing economic environment. Here we can make a distinction between adaptive agents and learning agents: the first use simple stimulus-response behaviour to only adapt their response to their environment, while the last use a conscious effort to learn about the underlying structure of their environment.

In the following more details are given regarding each market present in the model.

2.1 Goods and Labor Markets

For detailed information about the economic modelling choices characterizing the goods and the labor markets, see [15, 16]. See also [5] and [6] for additional explanations and for some discussion and analysis of computational experiments directly involving the two markets. What follows is a qualitative description of the main aspects that are relevant to the paper.

The goods markets are populated by IGFirms (investment goods firms) that sell capital goods to CGfirms (consumption goods firms), that produce the final consumption good. Stocks of firms product are kept in regional malls that sell them directly to households. A standard inventory rule is employed for managing the stock holding. Standard results from inventory theory suggest that the firm should choose its desired replenishment quantity for a mall according to its expectations on demand, calculated by means of a linear regression based on previous demands.

Consumption good producers need physical capital and labor to produce. The production technology in the consumption goods sector is represented by a Cobb-Douglas type production function with complementarities between the quality of the investment good and the specific skills of employees for using that type of technology. Factor productivity is determined by the minimum of the average quality of physical capital and the average level of relevant specific skills of the workers. Capital and labor input is substitutable with a constant elasticity and we assume constant returns to scale. The monthly realized profit of a consumption goods producer is the difference of sales revenues achieved in the malls during the previous period and costs as well as investments (i.e. labor costs and capital good investments) borne for production in the current period. Wages for the full month are paid to all workers at the day when the firm updates its labor force. Investment goods are paid at the day when they are delivered.

Consumption good producers employ a standard approach from the management literature, the so-called ‘break-even analysis’ to set their prices. The break-even formula determines at what point the change in sales becomes large enough to make a price reduction profitable and at what point the decrease in sales becomes small enough to justify a rise in the price. Basically, this managerial pricing rule corresponds to standard elasticity based pricing.

Once a month households receive their income. Depending on the available cash, that is the current income from factor markets (i.e. labor income and dividends) plus assets carried over from the previous period, the household sets the budget which it will spend for consumption and consequently determines the remaining part which is saved. This decision is taken according to the buffer-stock saving theory [7, 4].

At the weekly visit to the mall in his region each consumer collects information about the range of goods provided and about the prices and inventories of the different goods. In the Marketing literature it is standard to describe individual consumption decisions using logit models. These models represent the stochastic influence of factors not explicitly modelled on consumption.We assume that a consumer’s decision about which good to buy is random, where purchasing probabilities are based on the values he attaches to the different choices he is aware of. Since in our setup there are no quality differences between consumer goods and we also do not take account of horizontal product differentiation, choice probabilities depend solely on prices.

The labor market is governed by a matching procedure that relates directly workers looking for a job and firms looking for labor force. On the demand side, firms post vacancies with corresponding wage offers. On the supply side, unemployed workers or workers seeking for a better job, compare the wage offers with their actual reservation wages. Then the matching algorithm operates by means of ranking procedures on the side both of firms and households (see [10] for more details).

The algorithm might lead to rationing of firms on the labor market and therefore to deviations of actual output quantities from the planned quantities. In such a case the quantities delivered by the consumption good producer to the malls is reduced proportionally. This results in lower stock levels and therefore it generally increases the expected planned production quantities in the following period.

2.2 Credit and Financial Markets

For more detailed information on the financial market, see [13] and [14]. [19] shows also economic results obtained by means of computational experiments in the financial market, mainly regarding the problem of the equity premium puzzle.

The EURACE artificial financial market operates on a daily basis and is characterized by a clearing house mechanism for price formation which is based on the matching of the demand and supply curves. The trading activity regards both stock and government bonds, while market participants are households, firms and the governments. Both firms and governments may occasionally participate to the market as sellers, with the purpose to raise funds by the issue of new shares or governments bonds. Households provide most of the trading activity in the market, to which they participate both for saving and speculation opportunities. Household preferences are designed taking into account the psychological findings emerged in the framework of behavioral finance and in particular of prospect theory [22]. Households portfolio allocation is then modeled according to a preference structure based on a key prospect theory insight, i.e., the myopic loss aversion, which depends on the limited foresight capabilities characterizing humans when forming beliefs about financial returns (see [2]).

Firms finance investments and production plans preferably with internal resources. When these funds are not sufficient, firms rely on external financing, applying for loans to the banks in the Credit Market. The decision about firms loan request is taken by the bank to which the firm applies and depends on the total amount of risk the bank is exposed to, as increased by the risk generated by the additional loan. If a firm is credit-rationed in the Credit Market, then it has other possibilities of financing, i.e. issuing new equity on the financial market.

Commercial banks have two roles: one consists in financing the production activities of the firms, operating under a Basel II-like regulatory regime. The other role is to ensure the functioning of the payment system among trading agents. Finally, firms and households deposit entirely their liquid assets in the banks.

In the model banks are at the core of the system of payments: each transaction passes through the bank channel. Firms and households do not hold money as currency but under the form of bank deposits. Hence, the sum of payment accounts of bank's clients is equal to bank's deposits. As a consequence, every transaction (payment) between two non-financial agents translates into a transaction between two banks. At the end of every day, agents communicate the consistency of their liquid assets to their banks; then each bank can account for the net difference between inflows and outflows of money from and to the other banks and, if its reserves are negative, a compensating lending of last resort by the central bank is always granted. Thus, a sort of Deferred Net Settlement System has been implemented.

The Central Bank has several function in the EURACE model. It helps banks by providing them with liquidity when they are in short supply. It has the role of monitoring the banking sector setting the maximum level of leverage each bank can afford. It decides the lowest level of the interest rate, which is a reference value for the banking sector. If the quantitative easing feature is active, the central bank expands its balance sheet by purchasing government bonds in the financial market.

More details about the credit market of EURACE can be found in [11] and [12].

3 Results

Computational experiments has been performed in order to study the interplay between the supply of endogenous credit money and the performance of the economy, measured by the dynamics of the gross domestic product (GDP), the unemployment level, the dynamics of prices and the accumulation of physical capital in the economy.

The dynamics of credit money is fully endogenous and depends on the supply of credit from the banking system, which is constrained by its equity base, and the demand of credit from firms in order to finance their production activity. Alternative dynamic paths for credit money have been produced by setting different firms' dividend policies. The ratio d of net earnings that firms pay out as dividends has been exogenously set to four different values, namely, 0.6, 0.7, 0.8, and 0.9. It is clear that for higher values of d, firms' investments and hiring of new labor force must be financed more by new loans than by retained earnings, thus determining a higher amount of credit money in the economy.

Besides, the non conventional monetary policy practice called quantitative easing is considered, alongside the fiscal policy pursued by the Government. The central bank policy rate is kept fixed at low levels, however, if the quantitative easing (QE) policy is active, the central bank may buy government bonds directly on the market, thus increasing the overall amount of fiat money in the economy. Under quantitative easing, the government budget deficit is funded just by the issue and sale of bonds on the market. In this case, the intervention of the central bank is finalized to sustain bond prices and thus to facilitate the financing of government debt. If quantitative

Table 1 Values report the ensemble averages over three different simulation runs of mean monthly rates. Each run is characterized by a different random seed. For each simulation run, mean monthly rates are computed over the entire simulation period, except for the first 12 months which have been considered as a transient and discarded

d	QE	physical capital growth rate (%)	unemployment rate (%)	real GDP growth rate (%)
0.6	no	-0.0015	14.9	0.13
	yes	0.0021	12.3	0.14
0.7	no	0.0005	13.2	0.11
	yes	0.0027	11.4	0.15
0.8	no	0.0038	13.6	0.15
	yes	0.0035	11.8	0.15
0.9	no	0.014	10.8	0.17
	yes	0.013	11.7	0.16

easing is not active, the government budget deficit in funded both by the issue of new bonds in the market and by an increase of tax rates.

Each parameters' setting is then characterized by one of the four values of d and by a binary variable which denotes whether the quantitative monetary policy is adopted. The total number of parameters settings then sums up to 8. In order to corroborate the significance of results, for each parameters setting, three different simulation runs have been considered, where each run is characterized by a proper seed of the pseudorandom numbers generator. The same set of three random seeds has been employed for all parameters' settings.

The agents' population is constituted by 1000 households, 10 consumption goods producing firms, 1 investment goods producing firms, 2 banks, 1 government and 1 central bank. The duration of each simulation is set to 360 months (30 years).

Tables 1 and 2 report the simulation results for the main real and nominal variables of the economy, respectively, obtained with the 8 parameters' settings considered. Figures 1 and 2 show two representative time paths for some of the variables

Table 2 Values report the ensemble averages over three different simulation runs of mean monthly rates. Each run is characterized by a different random seed. For each simulation run, mean monthly rates are computed over the entire simulation period, except for the first 12 months which have been considered as a transient and discarded

d	QE	Private sector money endowment growth rate (%)	price inflation rate (%)	wage inflation rate (%)
0.6	no	-0.10	0.0019	0.19
	yes	-0.11	-0.0093	0.17
0.7	no	-0.09	-0.005	0.17
	yes	-0.11	-0.011	0.17
0.8	no	0.04	-0.08	0.18
	yes	0.089	0.030	0.18
0.9	no	0.35	0.12	0.28
	yes	0.33	0.10	0.27

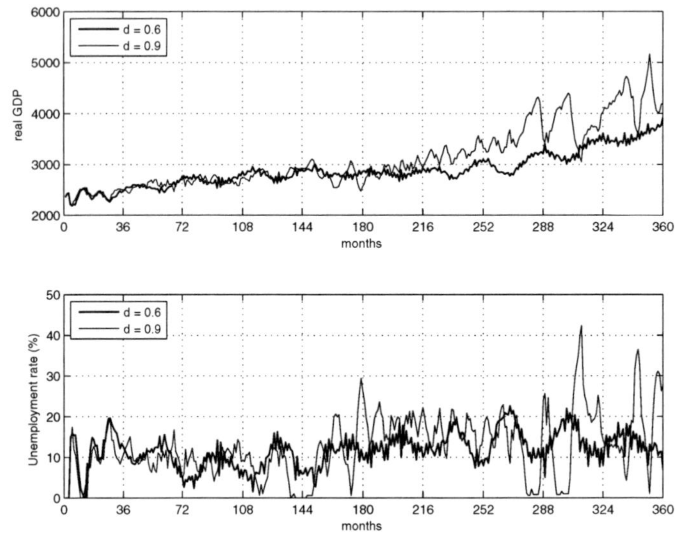

Fig. 1 Results of a simulation path for the real gross domestic product (GDP) and the unemployment rate. Two values of d are considered, i.e., $d = 0.6$ (thick line) and $d = 0.9$ (thin line)

considered. A clear and important empirical evidence that emerges from the path of GDP is that the EURACE model is able to exhibit endogenous business cycles. The main source of the observed business cycles is the strict relation between the real economic activity and its financing through the credit market, as it will be clear in what will follows.

Here, we will describe the simulation results with respect to the different values of d considered. The qualitative considerations which emerge with respect to the value of d do not depend whether the quantitative easing policy is active or not. In particular, Table 2 show that, as expected, an increase of the fraction d of firms' earnings, which is payed out as dividends, increases the private sector money endowment. The effects on nominal variables is also evident from the Fig. 2, where the simulation paths for the two extremes values of d, i.e., $d = 0.6$ and $d = 0.9$ are reported. The credit money supplied by the banking system is the source, together with the fiat money supplied by the central bank, of the endowment of liquid resources held by both the private sector (households, firms and banks) and the public sector (government and central bank). An increase in the demand for credit by firms, if supplied by banks, then increases the amount of liquid resources in the economy.

Table 2 also shows that higher inflation and wage rates are associated to higher values of d. Higher inflation rates for higher values of d should not be explained in this framework according to the quantity theory of money, due to the higher amount of liquidity in the economy. This because prices are not set by a fictitious Walrasian auctioneer at the cross between demand and supply, but are set by firms,

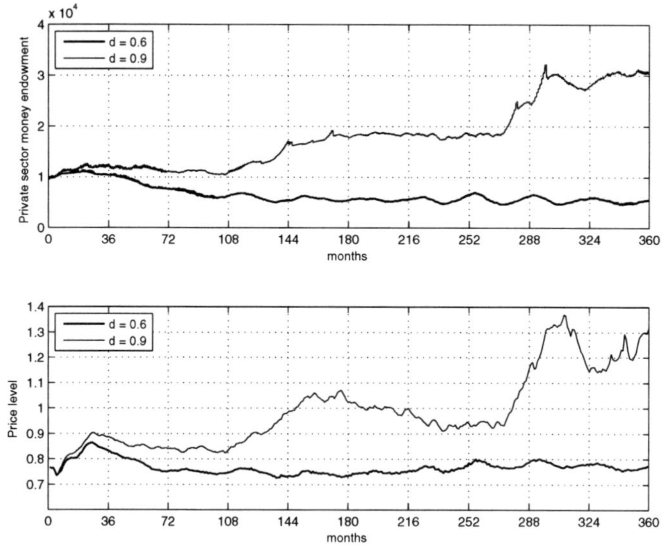

Fig. 2 Results of a simulation path for the private sector money endowment and the price level. Two values of d are considered, i.e., $d = 0.6$ (thick line) and $d = 0.9$ (thin line)

based on their costs, which are labor costs, capital costs and debt financing costs. Higher credit money means higher debt and higher debt financing costs. Higher credit money induces also higher wage inflation, and thus again higher price inflation through the cost channel. The wage inflation can be explained by the labor market conditions, i.e., the level of unemployment, as it will be clear in the following.

Table 1 presents the outcomes of the simulation concerning the real variables of the economy, i.e., unemployment level and rate of growth of physical capital and of real GDP. A clear indication emerges for a better macroeconomic performance, i.e., lower unemployment, and higher growth rate of real GDP and physical capital, related to higher levels of credit money in the economy. This indication is also corroborated by Fig. 1, where two simulation paths for the real GDP and the unemployment levels are reported for the two extreme values of d, i.e., $d = 0.6$ and $d = 0.9$.

It is worth noting, however, that a higher credit money may increases the amplitude of the business cycle. This feature emerges from Fig. 1 where both the real GDP and the unemployment level associated to $d = 0.9$ exhibit fluctuations which are far larger than the $d = 0.6$ case. In fact, a higher amount of credit money in the economy means higher levels of debt and leverage for firms. Firms bankruptcies due to insolvency (the equity goes negative) become more likely, and a firm bankruptcy causes mass layoffs and a sudden decrease in production. Besides, when a firm defaults on its debt, the lending bank suffers a reduction of its equity base, this in turn

Table 3 Values report the ensemble averages over three different simulation runs of the maximum percentage variability of the real GDP over a moving window of 60 month (5 years)

d	QE	first half (months: 12-180)	second half (months: 181-360)
0.6	no	-19.2	-18.7
	yes	-18.8	-22.0
0.7	no	-18.9	-16.6
	yes	-20.0	-21.0
0.8	no	-19.8	-26.1
	yes	-21.3	-22.9
0.9	no	-22.7	-31.1
	yes	-25.2	-37.5

determine a reduction of the supply of credit and the production sector may face a credit rationing, thus triggering further bankruptcies, due to liquidity problems.

Table 3, which reports the maximum percentage variability of real GDP in a moving time window of 60 months (5 years), confirms the graphic evidence of Fig. 1. In particular, a further interesting pattern emerges if we divide the time period of the simulation into two parts, and the maximum is computed separately in each part. For values of d like 0.8 and 0.9, i.e., a parameters setting where firms are more constrained to borrow credit money to fund their activity, the percentage variability in the second part of the simulation time span is clearly higher. This fact can be explained by looking to the leverage of firms and number of bankruptcies which is higher in the second part, after the first part is characterized by increasing levels of firms debt and leverage to unsustainable levels. This empirical finding may somewhat resembles the relation between the so-called "great moderation" period of the 90s and the first part of the 00s for the world economy, and the so-called "great recession" observed during the last two years.

The disaggregation of results with respect to the adoption or not of the quantitative easing monetary policy does not give a clear answer in the experiments considered about the goodness of a choice respect to another. The reason may be that in the setting considered, the government finances are usually sounds, so even if a quantitative easing policy is in principle adopted, it is applied just a few times.

4 Concluding Remarks

In this paper, we investigated the relationship between the amount of credit money and the macroeconomic performance in the EURACE simulator. Given that the dynamics of credit money is determined endogenously in the system, different dynamic paths have been produced by setting different firms' dividend policies. Results show the emergence of endogenous business cycles which are mainly due to the interplay between the real economic activity and its financing through the credit market. In particular, the amplitude of the business cycles strongly raises when the fraction of earnings paid out by firms as dividends is higher, that is when firms are more constrained to borrow credit money to fund their activity. This evidence can be explained by the fact that the level of firms leverage, defined as the debt-equity ratio, can be considered ad a proxy of the likelihood of bankruptcy, an event which causes mass layoffs and supply decrease.

Finally, this results show the possibility to explain the emergence of business cycles based on the complex internal functioning of the economy, without any ad-hoc exogenous shocks. The adopted agent-based framework has been able to address this complexity, and these results reinforce the validity of the EURACE model and simulator for future research in economics.

Acknowledgements This work was carried out in conjunction with the EURACE project (EU IST FP6 STREP grant: 035086) which is a collaboration lead by S. Cincotti (Università di Genova), H Dawid (Universitaet Bielefeld), C. Deissenberg (Université de la Méditerranée), K. Erkan (TUBITAK-UEKAE National Research Institute of Electronics and Cryptology), M. Gallegati (Università Politecnica delle Marche), M. Holcombe (University of Sheffield), M. Marchesi (Università di Cagliari), C. Greenough (Science and Technology Facilities Council Rutherford Appleton Laboratory).

References

1. Sallans, B., Pfister, A., Karatzoglou, A., Dorffner, G. (2003) Simulations and validation of an integrated markets model. Journal of Artificial Societies and Social Simulation **6** (4).
2. Benartzi, S., Thaler, R. H. (1995) Myopic loss aversion and the equity premium puzzle. The Quarterly Journal of Economics **110** (1): 73–92.
3. Bruun, C. (1999) Agent-based Keynesian economics: simulating a monetary production system bottom-up. University of Aaborg.
4. Carroll, C. D., 2001. A theory of the consumption function, with and without liquidity constraints. Journal of Economic Perspectives 15 (3): 23–45.
5. Dawid, H., Gemkow, S., Harting, P., Kabus, K., Neugart, M., Wersching, K. (2008) Skills, innovation, and growth: An agent-based policy analysis. Journal of Economics and Statistics **228** (2+3): 251–275.
6. Dawid, H., Gemkow, S., Harting, P., Neugart, M. (2009) On the effects of skill upgrading in the presence of spatial labor market frictions: an agent-based analysis of spatial policy design. Journal of Artificial Societies and Social Simulation **12** (4).
7. Deaton, A. (1992) Household saving in ldcs: credit markets, insurance and welfare. The Scandinavian Journal of Economics **94** (2): 253–273.
8. Deissenberg, C., van der Hoog, S., H., Dawid (2008) Eurace: A massively parallel agent-based model of the european economy. Applied Mathematics and Computation **204**: 541–552.
9. E. Kutschinski, T. U., Polani, D., September (2003) Learning competitive pricing strategies by multi-agent reinforcement learning. J Econometrics **27** (11-12): 2207–2218
10. Eurace (2009) Final activity report. `http://www.eurace.org/`
11. Eurace Project D5.1 (2007) Agent based models of goods, labour and credit markets. `http://www.eurace.org`
12. Eurace Project D5.2 (2008) Computational experiments of policy design on goods, labour and credit markets. `http://www.eurace.org`
13. Eurace Project D6.1 (2007) Agent based models of financial markets. `http://www.eurace.org`

14. Eurace Project D6.2 (2008) Computational experiments of policy design on financial markets. http://www.eurace.org
15. Eurace Project D7.1 (2007) Agent based models for skill dynamics and innovation. http://www.eurace.org
16. Eurace Project D7.2 (2008) Computational experiments of policy design on skill dynamics and innovation. http://www.eurace.org
17. LeBaron, B. D. (2006) Agent-based computational finance. Vol. 2 of Handbook of Computational Economics. North Holland
18. Tassier, T. (2001) Emerging small-world referral networks in evolutionary labor markets. IEEE Transaction of Evolutionary Computation **5** (5), 482–492
19. Teglio, A., Raberto, M., Cincotti, S. (2009) Explaining equity excess return by means of an agent-based financial market. Lecture Notes in Economics and Mathematical Systems. Springer Verlag, Ch. 12, pp. 145–156
20. Tesfatsion, L. (2001) Structure, behaviour, and market power in an evolutionary labour market with adaptive search. Journal of Economics Dynamics and Control **25**: 419–457
21. Tesfatsion, L., Judd, K. (2006) Agent-Based Computational Economics. Vol. 2 of Handbook of Computational Economics. North Holland
22. Tversky, A., Kahneman, D., October (1992) Advances in prospect theory: cumulative representation of uncertainty. Journal of Risk and Uncertainty **5** (4), 297–323

Reinforcement Learning of Heterogeneous Private Agents in a Macroeconomic Policy Game

Mahdi Hemmati, Masoud Nili, and Nasser Sadati

Abstract A repeated inflation-unemployment game within the linear-quadratic framework of Barro and Gordon is studied assuming that the government would like to cheat optimally and the finite heterogeneous population of private agents attempts to learn the government's targets using a reinforcement learning algorithm. Private agents are heterogeneous in their initial expectations of inflation rate but are assumed to utilize an identical anticipatory reinforcement learning process, namely Q-learning. In our heterogeneous setting, the only way for the private agents to achieve a zero value for their loss function, is for all of them to correctly anticipate the Nash equilibrium. It is of particular significance that such a solution requires a convergence of expectations across an initially heterogeneous population. Computer simulations have been conducted using different tuning parameters to investigate the convergence of our proposed model of learning process to Nash equilibrium.

1 Introduction

Learning in games has attained much attention in recent decade. Game theory has focused mainly on equilibrium concepts. Rational players have commonly known identical beliefs in equilibrium and by definition it is a self-enforcing state; once equilibrium is reached no player has incentives to leave the action or probability

Mahdi Hemmati
Intelligent Systems Lab, Department of Electrical Engineering, Sharif University of Technology, Azadi Ave., Tehran, Iran. e-mail: `hemmati@ee.sharif.edu`

Masoud Nili
Graduate School of Management and Economics, Sharif University of Technology, Azadi Ave., Tehran, Iran. e-mail: `m.nili@sharif.edu`

Nasser Sadati
Intelligent Systems Lab, Department of Electrical Engineering, Sharif University of Technology, Azadi Ave., Tehran, Iran. e-mail: `sadati@sharif.edu`

M. LiCalzi et al. (eds.), *Progress in Artificial Economics*, Lecture Notes in Economics and Mathematical Systems 645, DOI 10.1007/978-3-642-13947-5_18,

mixture over actions prescribed by the equilibrium strategy profile. However, equilibrium concepts do not explain how rational players get to have identical beliefs or, in other words, how this self-enforcing state arises.

A traditional explanation of equilibrium is that it results from the analysis, introspection and reasoning by the players in a situation where the rules of the game, the rationality of the players, and the players' payoff functions are all common knowledge. However, as mentioned in [1], these theories have many problems. First, a conceptual problem occurs when there are multiple equilibria since there is no explanation of how players come to expect the same equilibrium. Second, the hypothesis of exact common knowledge of payoffs and rationality might be arguable in many games. And third, equilibrium theory does a poor job explaining play in early rounds of most experiments, although it does much better job in later rounds.

Learning models on the contrary, explain how people learn, adapt or evolve toward equilibrium. [2] defines learning as an observed change in behavior owing to experience. However, it would be misleading if we describe learning models as models that explain how people learn or evolve toward equilibrium. This explanation assumes that players always learn the equilibrium and that the learning models converge always to equilibrium, which neither of them seems to be true. On the one hand, there are many experiments in which people deviate from the equilibrium predictions, such as the ultimatum mini-game experiments or many of the games in which there is a unique mixed Nash equilibrium. On the other hand, the learning models do not always converge. Different learning models have different convergence and stability properties and often these properties depend on the properties of specific games.

In this study we investigate whether Q-learning private agents in a macroeconomic game are able to learn and converge to Nash equilibrium. It is noteworthy that in many macroeconomic games, a population of agents plays against a single player, i.e., the government. Learning behavior of a population of agents is obviously much more complex than that of an individual player. Here, we try to gain some insights into this analytically intractable subject by means of multi-agent computer simulations. Moreover, various learning models differ in their informational and processing requirements and their assumptions about the rationality of agents. Another objective of this research is to find simple and less demanding learning models which lead to an optimal behavior. A side contribution of this paper is also taking the first steps toward bridging the gap between two isolated foundations of learning literature in economics and artificial intelligence disciplines.

The remainder of this paper is organized as follows. First, in Sect. 2 the macroeconomic policy game of study is explained. Specifically, incorporation of heterogeneous agents into the classic model would be elaborated. Section 3 is devoted to modeling of learning behavior of private agents. After a brief overview of reinforcement learning, the structure and mechanisms of Q-learning agents will be introduced. Simulation results are presented in Sect. 4. Finally, Sect. 5 restates the main results of this study and concludes the paper.

2 Inflation-Unemployment Game with Heterogeneous Agents

In this section, the macroeconomic policy game of interest is described. First we will give details of the repeated inflation-unemployment policy game with numerous homogeneous agents and then incorporation of heterogeneous agents into the model would be elaborated.

2.1 The Model with Homogeneous Agents

In a standard inflation-unemployment game, such as the linear-quadratic framework proposed by Barro and Gordon [3, 4], there are two players: the government and the private agents. They are leader and followers respectively. The government's instruments are both the announced and the actual rates of inflation, respectively π^a and π, while the private agents' instruments are their levels of expected inflation which are assumed to have an identical value of π^e. The instruments, π and π^e, critically determine the level of unemployment, u, assuming a standard Phillips curve relation:

$$u = u_n - c(\pi - \pi^e) \tag{1}$$

Here, u_n is the natural rate of unemployment, while c is a positive constant. In such a framework, it is possible for the government to drive the economy below the natural rate of unemployment, provided it can fool the private agents' expectations regarding inflation. One way to do so is to manipulate the announced inflation target, π^a, which in turn can impact the private agents' expectations. Consequently, the divergences between π and π^a corresponds to the willingness of the government to cheat. In a repeated-game, if the government cheats in a given period, $\pi \neq \pi^a$, the private agents may subsequently not believe such an announcement, so that there is an associated loss of credibility. Yet, this implies that the anticipations of the private agents must be defined by a reactive rule, which takes into consideration the discrepancies between announced and actual rate of inflation. If the private agents cannot surmise the government's targets, which are defined by a loss function, then their expectations may never converge to the true rate of inflation, π. Accordingly, the process of learning about the government's loss function can itself modify the evolution of the inflation path.

Let us assume that the government's loss function is given by a standard linear quadratic form:

$$J^L = \frac{1}{2}\left\{(u - \bar{u})^2 + (\pi - \bar{\pi})^2\right\} \tag{2}$$

where $\bar{u}$ and $\bar{\pi}$ are the targeted levels of unemployment and inflation respectively and u is determined by the Phillips curve relation in (1), once the government has set the inflation rate. It will be assumed that the private agents seek to minimize the errors in their expectations, as specified by the following loss function:

$$J^F = \frac{1}{2}(\pi^e - \pi)^2 \tag{3}$$

If the government is credible, then the actual rate of inflation is supposed to be identical to the announced one such that, $\pi = \pi^a$.

In light of Eqs. (1) and (2), the reaction function of the government is defined by minimizing its loss function with respect to its instrument, the inflation rate, such that:

$$\pi = T^L(\pi^e) \equiv \arg\min_{\pi} J^L(\pi, \pi^e) = \frac{c(u_n - \bar{u}) + c^2\pi^e + \bar{\pi}}{1 + c^2}. \tag{4}$$

Analogously, the reaction function corresponding to the private agents, is defined by minimizing the loss function (3) with respect to their expected inflation rate, so that:

$$\pi^e = T^F(\pi^a) \equiv \arg\min_{\pi^e} J^F(\pi^a, \pi^e) = \pi^a. \tag{5}$$

A one-period game can then be simulated for specific parameters and target values under alternative formulations for the government's conceivable strategies. More specifically, for a numerical example with $c = 1$, $\bar{\pi} = 2$, $\bar{u} = 5$ and $u_n = 7$ four different strategies are highlighted by the results in Table 1 [5]. Here the government alternatively makes a clear commitment such that the announced and actual rates of inflation are the same (case 1); whereby the government experiences a loss, but not the private agents. In cases 2 and 3, the government cheats, $\pi \neq \pi^a$, so that it eliminates its own losses, but this has the side-effect of generating high losses for the private agents. The final case corresponds to a Nash equilibrium in which, there is no cheating and hence no losses for the private sector.

Obviously, in case of a repeated game, cheating would lead the government to lose its credibility and hence the private agents have to learn the probable evolution of the government's inflation target. An optimal learning process should converge to the Nash inflation level, since at this point the expected and actual rates of inflation are the same.

2.2 Incorporation of Heterogeneous Agents

Here we assume that there exist N heterogeneous private agents with respect to their expected rates of inflation. In this case the aggregate anticipated rate of inflation in

Table 1 Numerical Results of One-Period Inflation-Unemployment Game

Strategies	π^a	π^e	π	J^L	J^F
(1) Commitment	2	2	2	2	0
(2) Cheating (No Commitment)	2	2	3	1	0.5
(3) Optimal Cheating	0	0	2	0	2
(4) Nash	4	4	4	4	0

period t becomes an arithmetic average of the individuals' different expectations, which is assumed to be known by the government:

$$\pi_t^e = \frac{1}{N}\sum_{i=1}^{N}\pi_{i,t}^e. \tag{6}$$

We continue to assume that each private agent i seek to minimize the errors in its expectations:

$$J_i^F\left(\pi_t, \pi_{i,t}^e\right) = \frac{1}{2}\left(\pi_t - \pi_{i,t}^e\right)^2. \tag{7}$$

As in the previous one-period example, assuming that the government wants to cheat optimally the private agents, then the government will announce in the first period a zero rate of inflation, $\pi_1^e = 0$. Given such an announcement, each agent will choose a level of expected inflation, $\pi_{i,1}^e$, with $i = 1,...,N$. This will depend on its assessment of the government's credibility. The average anticipated value of inflation in the first period, $\pi_1^e = \frac{1}{N}\sum\pi_{i,1}^e$, will be influenced by the heterogeneity of the agents' expectations, where $\pi_{i,1}^e \neq \pi_{j,1}^e$ for some agents i and j. In light of this expected inflation rate, π_1^e, the government is free to choose the actual rate, π_1, where the reaction function in (4) again applies.

For those private agents whom are fooled in period one, the credibility of the government's subsequent announcement in period two is lost. As a consequence, each individual private agent will seek to learn the government's targeted inflation rate. This leads to a dynamic game, where the private agents' loss is minimized by the Nash inflation target value. That is, each private agent, by choosing its expected rate, will impact the average level of expectations to which the government will, in turn, respond by setting an optimal inflation rate in the same period.

In summary, the following decision sequence takes place within the framework of our repeated macroeconomic policy game: (1) Agent i forms her expectation by choosing $\pi_{i,t}^e$, which then would result in an average expected inflation rate, π_t^e; (2) the government reacts to the average of expectations by calculating and applying the actual inflation rate, $\pi_t = T^L\left(\pi^e\right)$.

The only way for the private agents to achieve a zero value for their loss function, is to anticipate the Nash equilibrium. It is of particular significance that such a solution requires the convergence of expectations across an initially heterogeneous population.

In our numerical example, it is when $\pi_t^e = 4$ and therefore $\pi_t = 4$ which results in $J_i^F = 0, \forall i$. That is, the optimal learning process should lead the private agents, as a whole, to set the average anticipated rate of inflation to the Nash value. However, perfect stability of the learning process will be reached only if $\pi_{i,t}^e = 4, \forall i$. Otherwise, some agents will have an incentive to change their anticipations.

3 Modeling of Learning Private Agents

The literature on modeling the learning behavior of economic agents is quite rich. Overviews of this literature are provided by [6, 7]. One can distinguish between *individual learning* models and *social learning* models [8]. In individual learning models an agent learns exclusively from its own experience, whereas in social learning models an agent also learns from the experience of other agents.

The two most important approaches to modeling individual learning behavior are *belief-based learning* and *reinforcement learning*. Examples of belief-based learning models are Cournot adjustment and fictitious play [1]. These two models assume that an agent has the ability both to observe its opponents' action choices and to calculate best responses. In a Cournot oligopoly game, the models predict that firm behavior can converge only to the Nash equilibrium. Reinforcement learning is based on a very simple idea: the higher the payoffs obtained from a strategy in the past, the more likely the strategy is to be played.

3.1 Why Reinforcement Learning?

In this study, reinforcement learning [9] has been employed to model the learning behavior of private agents. Compared with belief-based learning models, reinforcement learning models make few assumptions about both the information available to an agent and the cognitive abilities or rationality of an agent. For example, in reinforcement learning an agent needs no information about its opponents' action choices or about the payoffs of the game. An agent is only assumed to have knowledge of the strategies that it can play and, after playing a strategy, of the payoff that it has obtained from that strategy. Reinforcement learning models, also called stimulus-response, have their roots in behaviorist psychology, dated back to 1920. After behaviorist introduced reinforcement learning, Bush and Mosteller [20] formalized simple reinforcement rules and applied them to learning in decisions. Cross [10, 11] applied reinforcement learning to economic decisions. Finally in the nineties, reinforcement learning was revived; first Arthur [12, 13] applied it to the behavior of economic agents and later Mookerjee and Sopher [14], Roth and Erev [15, 16] and Sarin and Vahid [17] applied reinforcement to games.

An extensive bulk of research has shown that reinforcement learning agents benefit from a high reproduction capability of human-like behaviors [12, 13, 15, 16]. For example, Roth and Erev compared simulation results of simple reinforcement learning agents with results of subject experiments in several examples [15, 16] revealing that: (1) computer simulation using simple reinforcement learning agents can better explain the result of subject experiments than economic theory; and (2) the former approach has greater potential of predicting results than the latter.

Reinforcement learning models are studied both in the economic literature and in the artificial intelligence literature (for an overview of the artificial intelligence (AI) literature on reinforcement learning, see [18, 9]). The reinforcement learning agents, specifically, Q-learning agents [19] in computer science literature are employed among other famous agents like Roth and Erev's learning agent [15, 16] in social science literature. This is because previous research revealed that Q-learning agents can learn consistent behaviors and acquire sequential negotiation in the sequential bargaining game, while Roth and Erev's agents cannot. Roth and Erev's agents work well only in one-time negotiation.

3.2 Overview of Reinforcement Learning

The basic intuition underlying *reinforcement learning* (RL) is that the tendency to implement an action should be strengthened (reinforced) if it results in favorable outcomes and weakened if it results in unfavorable outcomes [9]. RL is a goal-directed learning from *interaction* of the *agent* with the *environment*. In other words, reinforcement learning is learning what to do -how to map situations to actions- so as to maximize a numerical reward. The learner is not told which actions to take, but instead must discover which actions yield the most reward by trying them.

In the most interesting and challenging cases, actions may affect not only the immediate reward but also the next situation and, through that, all subsequent rewards. These two characteristics, *trial-and-error search* and *delayed reward*, are the two most important distinguishing features of reinforcement learning. Actually, decision problems can be divided into two types: *non-sequential* and *sequential*. In non-sequential decision problems, an agent must learn a mapping from states to actions that maximizes expected immediate reward. In sequential decision problems, the agent must again learn a mapping from states to actions, but now the actions selected by the agent may influence future situations and hence future rewards as well as immediate rewards. Consequently, it might be advantageous for the agent to engage in anticipatory evaluation of the future possible consequences of its current actions.

In the early RL literature, the focus was generally on a single agent playing a game against nature. That is, a single agent was assumed to face a decision problem in which his uncertainty regarding which action to choose arose from an exogenous source modeled as a probability distribution that was independent of the agent's action choices. Recently, theoretical game theorists have begun to explore the use of RL in multi-agent contexts as a "selection principle" for determining the selection of a particular Nash equilibrium when multiple Nash equilibria exist [1]. Also, RL methods are being used to explain experimental data obtained from human subjects who are learning to play repeated games; see [15, 16]. Learning among multiple strategically interacting agents is far more difficult to study than learning in games against nature since the choice environment of each agent is now intrinsically non-stationary.

A key aspect of all learning models/algorithms is the amount of *anticipation* (look-ahead) that agents employ. At one extreme, an agent might rely entirely on *reactive (stimulus-response) learning* for choosing its actions. A state-of-the-environment (stimulus) occurs, and the agent reacts to this state by choosing a particular action (response). The agent then observes an outcome, and it uses this outcome to either weaken or strengthen the association between the state and the action in the future. In this scenario, then, the agent continuously adapts to its environment by asking the question: if this state occurs, what action should I take? However, the agent does not deliberately attempt to modify its environment to suit its own purposes. At the other extreme, in the *anticipatory learning* case, the agent deliberately attempts to modify its environment to suit its own purposes by asking the following forward-looking question: if I take this action now, what outcomes might occur in the future? The agent chooses its actions in an attempt to increase the likelihood that the ensuing outcomes will be favorable to itself.

One complication in the design of anticipatory learning methods is the well-known *dual decision problem* from adaptive control theory. Namely, from a longer-run point of view, it might be optimal for an agent to sacrifice short-term reward for information gain through experimentation, if this ultimately permits the agent to make more informed future decisions because it has better estimates of the true outcome probability distributions. This challenge is called the *trade-off between exploration and exploitation* in RL literature. To obtain a lot of reward, an anticipatory reinforcement learning agent must prefer actions that it has tried in the past and found to be effective in producing reward. But to discover such actions, it has to try actions that it has not selected before. The agent has to *exploit* what she already knows in order to obtain reward, but she also has to *explore* in order to make better action selections in the future.

3.3 Structure of our RL Agents

This section explains the structure and implementation of learning private agents of the economy modeled as anticipatory reinforcement learning agents. We assume that all stage games are indistinguishable to the agent. In other words, the agents are facing a *single-state game*. Therefore, at different stages of the game, agents remain at the same single state of the game.

Each agent is provided with a *memory* which is supposed to save and keep Q-values associated with different actions at his disposal, i.e. different levels of anticipated inflation. Without loss of generality, the interval of feasible anticipations is taken as $[0,10]$ and it is assumed that agents choose their anticipation among the set of discretized integer values: $0,1,...,10$. This means that each agent encounters 11 possible actions at each time period and saves an array of Q-values corresponding to these 11 possible actions. Agents update their array of Q-values using standard Q-learning algorithm and choose their anticipations using *ε-greedy action selection* method. ε is initially set to 1 to represent an untrained and inexperienced agent, resulting in full exploration in the early stages of the game. During the process of learning, ε is gradually decreased in each time step to reflect the increasing experience of the agent, making exploitation of previous experiences more probable.

4 Simulation Results

Computer simulations has been carried out to investigate the outcome of the explained macroeconomic policy game with *all* private agents modeled as simultaneous Q-learners in a multi-agent environment and the government as a rational decision maker who plays optimally. We focus on the long-run behavior of Q-learning agents and seek to find out whether the Nash equilibrium is attained. As previously mentioned, to reach Nash equilibrium of this game, it is necessary for *all* agents to learn the level of inflation at Nash equilibrium. This means that the population of agents which is assumed to be initially heterogeneous in terms of their initial anticipation of inflation, should move toward homogeneity of anticipations.

The following numerical values are used for the macroeconomic framework of the game: $c = 1$, $\bar{\pi} = 2$, $\bar{u} = 5$ and $u_n = 7$. As stated before, Nash equilibrium for these values would be $\pi^{Nash} = 4$. The population size or the number of private agents in the economy is taken to be $N = 100$. In the first period the government announces zero inflation. Then, agents are assumed to choose their initial anticipated inflation *randomly* from the interval of [0,5]. This is the major source of heterogeneity of private agents in our settings.

The repeated game is simulated for $T = 200$ time steps. Regarding the stochastic nature of reinforcement learning behavior of agents and also random distribution of their initial anticipations, numerous (50 to 300) runs of a single simulation has been taken to extract the statistical average of aggregate quantities, namely, π^e and π. In updating Q-values, we have assumed that $\gamma = 0$. This means that the agent is ignoring his future rewards, while making decision about his current action. Since our game setting is single-state and all stage games seem the same to the agents, it is reasonable to assume that agents would prefer to optimize only their current reward.

In order to investigate the long-run convergence of the population of heterogeneous agents to Nash equilibrium, we need to introduce a quantitative index or criterion of convergence for a batch of simulation runs with fixed parameters. First, we define *convergence error* of each simulation run as the absolute difference between the steady-state value of anticipated inflation and inflation level at Nash equilibrium:

$$\delta = \left| \pi_{ss}^{e} - \pi^{Nash} \right| . \tag{8}$$

Now, for set of simulation parameters, we define the *index of convergence with a tolerance of* Δ as the following ratio:

$$I_{\Delta} = \frac{\text{Number of Simulation Runs with } \delta \leq \Delta}{\text{Total Number of Simulation Runs}} . \tag{9}$$

Different values of learning rate, α, for Q-learning agents have been tried through a number of simulation runs. Simulation results for $\alpha = 0.75$, $\alpha = 0.8$ and $\alpha = 1.0$ are shown in Fig. 1 to Fig. 3, respectively. It shows that despite the multi-agent and nonstationary environment of the economy, the aggregate behavior of the private sector, based on reinforcement learning of private agents, is converging to Nash

Table 2 Index of Convergence for some simulation cases

Learning Rate	$I_{0.2}$	$I_{0.3}$	$I_{0.4}$
$\alpha = 0.75$	0.82	0.94	1.0
$\alpha = 0.8$	0.75	0.95	1.0
$\alpha = 1.0$	0.94	1.0	1.0

equilibrium. Quantitatively speaking, a few indices of convergence for these two cases are listed in Table 2. For example, $I_{0.2} = 0.82$ means that in %82 of simulation runs with $\alpha = 0.75$, the convergence error had been less than ± 0.2; that is the steady-state anticipated inflation would fall into the interval [3.8,4.2] around the Nash equilibrium, $\pi^{Nash} = 4$. The most important observation is that the best results is achieved when $\alpha = 1.0$. It means that when agents make their decisions only based on the reward of last stage, they are more likely to learn the Nash equilibrium.

5 Conclusion

In this paper the aggregate behavior of a population of RL private agents in a macroeconomic policy game was studied. This repeated game runs between the rational and optimally-behaving government, from one side and a number of heterogeneous private agents with a minimum level of rationality, information and processing capability assumed, from another side. Using agent-based simulations, we showed that Q-learning agents would learn the Nash equilibrium of the game and the aggregate behavior of the population would converge to the Nash equilibrium in a statistical sense.

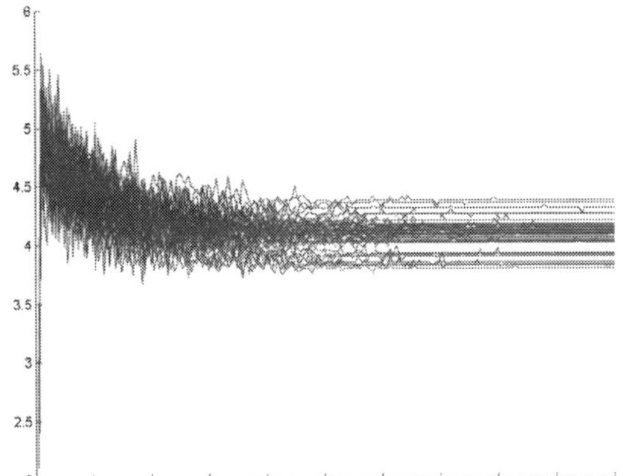

Fig. 1 Anticipated Inflation, π^e, for $\alpha = 0.75$ in 50 Simulation Runs

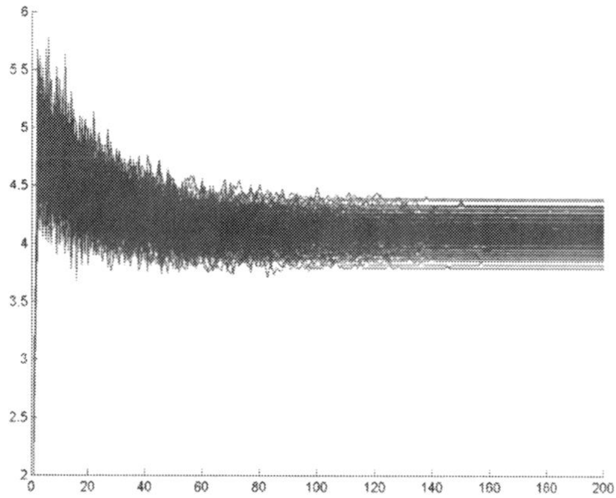

Fig. 2 Anticipated Inflation, π^e, for $\alpha = 0.8$ in 50 Simulation Runs

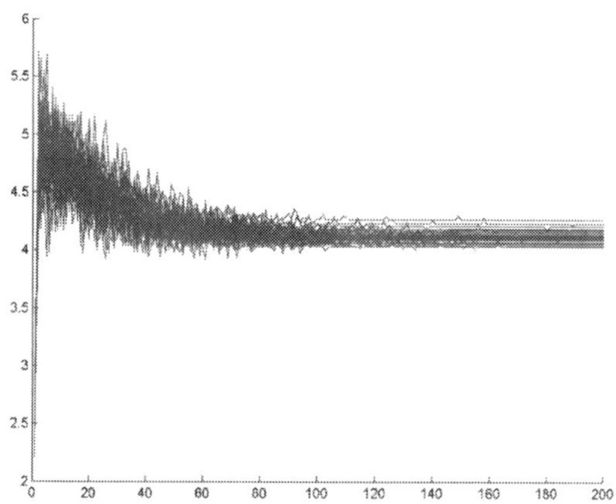

Fig. 3 Anticipated Inflation, π^e, for $\alpha = 1.0$ in 50 Simulation Runs

It should be noted that the underlying game of this study was a single-state repeated game, where agents could not differentiate between different stages. In order to exploit the potential capabilities of Q-learning as an anticipatory reinforcement learning method, it is needed to reformulate the studied macroeconomic game as a sequential and dynamic game with multiple states. This issue remains for future research.

References

1. Fudenberg, D., Levine, D.K. (1998) The Theory of Learning in Games. MIT Press, Cambridge, MA

2. Camerer, C. F. (2003) Behavioral Game Theory: experiments in strategic interaction. Princeton University Press, Princeton, NJ
3. Barro, R., Gordon, D. (1983) A Positive Theory of Monetary Policy in a Natural Rate Model. Journal of Political Economy **9**(4): 589–610
4. Barro, R., Gordon, D. (1983) Rules, Discretion and Reputation in a model of Monetary Policy. Journal of Monetary Economics **12**(1): 101–121
5. Vallée, T., Deissenberg, C., Basar, T. (1999) Optimal Open Loop Cheating in Dynamic Reversed Linear Quadratic Stackelberg Games. Annals of Operations Research **88**(1): 217–232
6. Brenner, T. (2006) Agent learning representation: advice on modelling economic learning. In: Tesfatsion, L., Judd, K.L. (eds.), Handbook of Computational Economics, vol. 2, pp. 895–947, Elsevier, Amsterdam
7. Duffy, J. (2006) Agent-based models and human subject experiments. In: Tesfatsion, L., Judd, K.L. (eds.), Handbook of Computational Economics, vol. 2, pp. 949–1011, Elsevier, Amsterdam
8. Vriend, N.J. (2000) An illustration of the essential difference between individual and social learning, and its consequences for computational analyses. Journal of Economic Dynamics and Control **24**: 1–19
9. Sutton, R.S., Barto, A.G. (1998) Reinforcement Learning: An Introduction. MIT Press, Cambridge, MA
10. Cross, J. G. (1973) A stochastic leaning model of economic behavior. Quarterly Journal of Economics **87**: 239–66
11. Cross, J. G. (1983) A theory of adaptive economic behavior. Cambridge Univ. Press, London
12. Arthur, B. (1991) Designing economic agents that act like human agents: A behavioral approach to bounded rationality. American Economic Review Proceedings **81**: 353–359
13. Arthur, B. (1994) On designing economic agents that behave like human agents. Journal of Evolutionary Economics **3**: 1–22
14. Mookherjee, D., Sopher, B. (1997) Learning and decision costs in experimental constant sum games. Games and Economic Behavior **19**: 97–132
15. Roth, A.E., Erev, I. (1995) Learning in extensive form games: experimental data and simple dynamic models in the intermediate term. Games and Economic Behavior **8**:164–212
16. Erev, I., Roth, A.E. (1998) Predicting how people play games: reinforcement learning in experimental games with unique, mixed strategy equilibria. American Economic Review **88**:848–881
17. Sarin, R., Vahid, F. (2001) Predicting how people play games: a simple dynamic model of choice. Games and Economic Behavior **34**: 104–122
18. Kaelbling, L.P., Littman, M.L., Moore, A.W. (1996) Reinforcement learning: a survey. Journal of Artificial Intelligence Research **4**: 237-285
19. Watkins, C.J.C.H., Dayan, P. (1992) Q-learning. Machine Learning **8**:279–292
20. Bush, R., R. Mosteller. (1955) Stochastic Models for Learning. John Wiley, New York, NY

Part VII
Demographics and Culture

Towards an Agent-Based Model of the Economic Development Process: The Dynamics of the Fertility Rate

Gianfranco Giulioni and Edgardo Bucciarelli

Abstract This paper is a first step to build an agent-based model of the economic development process. We focus on households' behavior by studying in particular the relationship between the available income and the optimal choice on quantity (fertility) and quality (level of education) of children. A collection of households taking decisions according to the rules identified at the individual level, but perturbed by idiosyncratic shocks and subject to a mean field interaction are monitored by using computer simulations. The model gives us the opportunity to investigate the evolution of the distributions of fertility and income by using data recorded from simulations at the individual level. Averaging the number of children across households we find that this model to be able to replicate the J-shaped pattern of the fertility rate found in recent empirical analysis.

1 Introduction

The relative inconsistency of exogenous and endogenous growth models with some of the most basic features of the economic development process has led recently to a search for a unified theory. The goal is that to discover the underlying micro-foundations of the growth dynamics in their entirety, capturing simultaneously the age of Malthusian stagnation that characterized most of human history, the contemporary era of modern economic growth and the fundamental driving forces that triggered the transition between these socio-economic systems and the associated phenomenon of the great divergence in income per capita across countries. The existing theoretical literature on the transition from stagnation to growth has drawn attention mostly to the role played by endogenous technological progress or human-capital accumulation. However a number of leading economists pointed out the importance

Gianfranco Giulioni and Edgardo Bucciarelli
Department of Quantitative Methods and Economic Theory, University of Chieti-Pescara, Italy
e-mail: g.giulioni@unich.it, e.bucciarelli@unich.it

M. LiCalzi et al. (eds.), *Progress in Artificial Economics*, Lecture Notes in Economics and Mathematical Systems 645, DOI 10.1007/978-3-642-13947-5_19,

of fertility choices in the development process. [9] notices that a feedback loop between technology and population generates a transition from the proximity of a Malthusian equilibrium to the post Malthusian regime. As hypothesized by [4, 3] and later by [11, 12], the increasing proportion of young adults would generate a downward pressure on young men's relative wages, or on the size of land holdings passed on from parent to child, leading them to reduce fertility in order to achieve their desired level of material aspirations. This in turn would cause young adults to accept a trade-off between family size and material well-being, in which total fertility rates tumble as social norms regarding individual control of fertility and acceptable family sizes begin to change. This seems to be an aspect of the demographic transition which has been overlooked in the past because of a focus on absolute rather than relative income [2] and this phenomenon is observed in country after country, above all in U.S., which has begun the fertility transition since 1950 and evidence suggests that this was the case in earlier transitions as well. From a more macroeconomic point of view, [10] introduces a Malthusian model in which households streamline over fertility and consumption. [6] and [7] study models where the transition from Malthusian stagnation to modern growth is a feature of the equilibrium growth path. The need for further work in this direction has been recently stated by [5].

The conventional belief that developed countries tend to have lower fertility rates while fertility rates in developing realities are high is grounded on real world observation. At first this means that declining fertility rates are attributed to advances in economic development. However, this negative relationship seems to be inverted when the level of development is relatively high. Recently [14] performed cross sectional and longitudinal analyses to examine the correlation between the total fertility rate (TFR) and the human development index (HDI), whereas their main finding is that in highly developed countries with HDI above 0.9, further development halts the declining fertility rates. This means that the previously negative development-fertility association is reversed and becomes J-shaped. There was occurred a fundamental change in the well-established negative relationship between fertility and development in the last few decades. This change not only challenges conventional knowledge but also would require that policy makers re-evaluate their present assumptions regarding the fertility-development relationship when they hammer out future policies.

The goal of this paper is to integrate the standard microeconomic analysis with new elements. In particular, we pick up the recent concept that one should incorporate increasing marginal utility of consumption to understand poverty traps [8]. In our model, agents give an increasing importance to consumption when they are poor. Then, consumption gradually looses importance as living conditions improve. This change of behavior also modifies people's motivation for having children. This pushes us to meet knowledge developed in other fields of sciences like sociology where it has been noted how children starting from being useful they become priceless over time [16].

We start our investigation by looking at the dynamics of fertility rates found in the reality in Sect. 2. We then proceed by providing a microeconomic foundation for an household's choice on number and education of children (Sect. 3). We first look at this problem analytically in Sect. 3.1. As we will explain below we have non-convexities in the individual problem, so that the internal maximum found by the analytical procedure could not be the global one. We numerically study the shape of the utility function and the solution of the maximization problem in Sect. 3.2. Section 4 is dedicated to the study of a system made up of a collection of households who take decisions according to the rules identified in Secs. 3.1 and 3.2. There we will show how the presence of idiosyncratic shocks and the externalities introduced by mean field interactions bring the simulation results towards a realistic outcome. In Sect. 5 we draw conclusions and discuss future developments.

2 The Evolution of the Total Fertility Rate in Developed Countries

It will be useful to have a term of comparison for the results of our simulations. To this aim we discuss the patterns observed in the time evolution of the TFR for a number of developed countries. Let us first define the TFR. It estimates the number of births which a hypothetical cohort of women would have if they experienced throughout their childbearing years the same age-specific birth rates observed for a cross-section of woman having different ages in a given year and if they were not subject to mortality. The rate can be expressed as the average number of children that would be born per woman. Because it is computed from age-specific birth rates, the TFR is age adjusted and can be readily be compared for populations across time or among geographic areas.

We report the TFR time series for the United States in Fig. 1. Throughout the nineteenth and the first part of the twentieth century, fertility rates in the United States declined until 1946 when rates start increasing in a significant way. Following World War II, fertility rates among American women in childbearing age increased in 1945 and 1946, and they remained unusually high for a while. After 1957, TFR declined until the mid 1960s. Referred to as the so-called "baby boom", this historic abnormality in U.S. fertility occurred between 1946 and 1964. Following the "baby boom", except for small increases in the later part of the 1960s into the early 1990s, fertility remained stable [13].[1] Thereabouts the same dynamics has been surveyed for the countries considered in Fig. 2. The figure shows the dynamics of TFR in France, Germany, Italy, United Kingdom and United States since World War II; the projections up to 2050 estimated by the United Nations are also displayed. In conclusion, the year 1996 based projections seem to suggest that the TFR will converge to a common value in developed countries. Countries whose TFRs are below this value, in particular Germany and Italy, will be characterized by a growing TFR in the future. Note that as long as a Country's development level increases with

[1] The population has grown throughout the century but at a declining rate. Between 1901 and 1911 the growth rate of the population averaged about 1% per annum. Between 1981 and 1991 the average growth rate of the population had fallen to about 0.25% per annum.

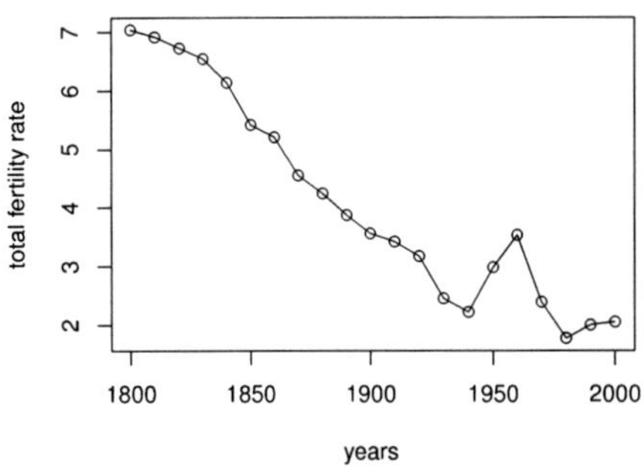

Fig. 1 United States' total fertility rates.[a]

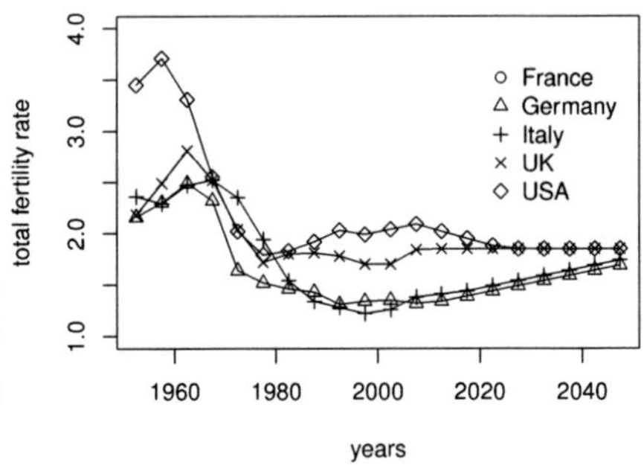

Fig. 2 total fertility rate for a number of developed countries.[b]

[a] data from `http://eh.net/encyclopedia/article/haines.demography`
[b] data from `http://esa.un.org/UNPP/index.asp?panel=2`.

time, the displayed forecast imply that the J-shaped fertility-development relationship found by [14] will become more evident for a number of countries.

3 The Household's Problem

3.1 Analytical Insights

In line with the widely accepted view in nowadays macroeconomics, we start by giving microfoundations to the agent's behavior. In this paper we focus on the household's problem. The decision on the number of children and the level of their education are important for us being this a first step towards an agent-based model of the development process. As in other models, each individual has one parent and its relevant (for the analysis) life is divided into two parts. In the first part she has no decision to take and this time span is dedicated to follow the parent's guidelines. In the second one, an individual earns an income that she allocates in her own consumption and expenditures for children. The parent's problem is that of maximizing utility under the balance sheet constraint. Quantity and quality of children are choice variables beside the level of consumption.[2] The utility function we use is similar to that introduced by [1] and the problem is that to maximize

[2] The models developed using these settings are referred to as overlapping generations models. However, in the model presented in this paper children have no active role, and the dynamics can be completely described in terms of the adults' variables.

$$U_{p,t} = c_t^{\alpha_U} + a_U \left(n_t E[y_k]\right)^{\beta_U}$$

subject to

$$y_t \geq c_t + n_t p_n + n_t p_e e_t$$
$$0 \leq n_t \leq \tilde{n}, \; e_t \geq 0 \text{ and } c_t \geq 0$$

where $U_{p,t}$ is the parent's utility, c_t is her consumption, n_t the number of children, $E[y_k]$ the expected income that will be earned by each kid, y_t the parent's income, e_t the level of education given to each child, p_e the cost of one unit of education, p_n the other cost to grow a child and $\tilde{n}$ is the maximum number of children allowed by nature. α_U, a_U and β_U are parameters. We assume the parent's expectation to be formulated as follows $E[y_k] = e_t^{\gamma}$, where γ measures the parent's confidence in education as a determinant of income.

The choice variables (n_t, e_t and c_t) can be determined by solving a system of three equations given by two first order conditions (these for e_t and n_t) and the constraint. A general analytical solution is not possible, however, in this section, we proceed by analyzing solvable cases.

At this point it is useful to discuss the meaning of the parameters. a_U and β_U regulate the household's behavior towards the future generation. In this paper we keep them constant requiring $a_U > 0$ and $0 < \beta_U < 1$.

The key parameters for the present study are γ and α_U. The latter is important especially for analytic purposes (γ, as we model it below, does not affect the analytic solution). Let us focus on α_U which determines the marginal utility behavior of the household's consumption. From a modeling point of view we assume there is a level of consumption ($\hat{c}_t$) greater than the subsistence one ($\tilde{c}_t$) such that $\alpha_U > 1$ when $c_t < \hat{c}$ and $\alpha_U < 1$ when $c_t > \hat{c}$. We focus on two cases which allow us to calculate analytical solutions: 1) $\alpha_U = 2 - \beta_U$ (for increasing marginal utility) and 2) $\alpha_U = \beta_U$ (for decreasing marginal utility).

In the first case the solution is

$$e^* = \frac{\gamma}{1-\gamma}\frac{p_n}{p_e}, \qquad n_t = \frac{B}{c_t}, \qquad c_t = \frac{y_t + \sqrt{y_t^2 - 4B(p_n + p_e e^*)}}{2}, \tag{1}$$

while in the second it is

$$e^* = \frac{\gamma}{1-\gamma}\frac{p_n}{p_e}, \qquad n_t = Bc_t, \qquad c_t = \frac{y_t}{(p_e e^* + p_n)B + 1}. \tag{2}$$

with

$$B = \left(\frac{\beta_U(1-\gamma)}{\gamma p_n}\right)^{\frac{1-\gamma\beta_U}{1-\beta_U}} \left(\frac{\beta_U}{p_e}\right)^{\frac{\gamma\beta_U}{1-\beta_U}} \left(\frac{a_U \gamma}{\alpha_U}\right)^{\frac{1}{1-\beta_U}}. \tag{3}$$

Case 1) has the potential to provide an explanation for the negative relationship between income and natality rate seen in real data. In fact from equations (1) we have $dc_t/dy_t > 0$, but $dn_t/dc_t < 0$ so that $dn_t/dy_t < 0$. In the second case there is a positive relationship between income and fertility.

3.2 Numerical Investigation

The need for a general knowledge of the solution gives us motivations to investigate the household's problem numerically. Firstly, due to the non-convexity of the utility function, the critical point found for the case $\alpha_U = 2 - \beta_U$ could not be the global maximum; in these cases the corner solution has to be considered and it is easy to compute it numerically. Secondly, in a numeric context, the decision on education, fertility and consumption can be obtained under the more general assumption of a smooth change in the parameters rather than for special values such the ones fixed in the previous section.

We let parameters γ and α_U to depend on the parent's income y_t:

$$\gamma_t = 1 - \frac{1}{y_t} \qquad \alpha_{U,t} = \alpha_{U\min} + \frac{1}{y_t}. \tag{4}$$

The solution for education is the one we have found in the previous section:

$$e_t^* = \frac{\gamma_t}{1-\gamma_t}\frac{p_n}{p_e}. \tag{5}$$

Substituting e_t^* in the utility function we obtain

$$U_{p,t} = c_t^{\alpha_{U,t}} + a_U\left[n_t(e_t^*)^{\gamma_t}\right]^{\beta_U},$$

and using the balance sheet constraint we arrive to the following expression:

$$U_{p,t} = (y_t - n_t p_n - n_t p_e e_t^*)^{\alpha_{U,t}} + a_U\left[n_t(e_t^*)^{\gamma_t}\right]^{\beta_U}. \tag{6}$$

This function has n_t as the only choice variable. We numerically determine the maximum value of this function in the interval $[0, \tilde{n}]$. We are particularly interested on how the maximum changes with the available income.

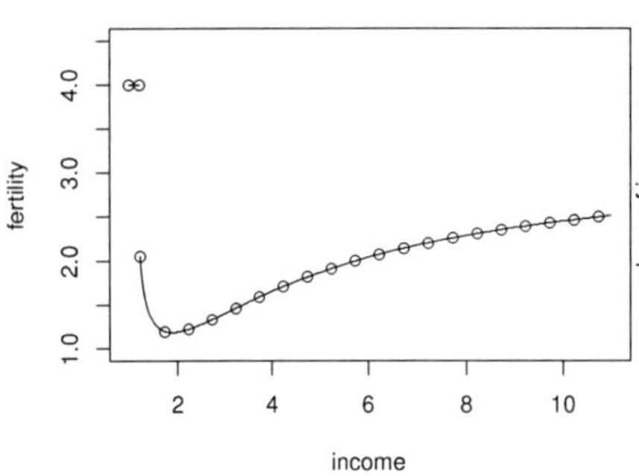

Fig. 3 Relationship between the fertility choice and income from the numeric solution

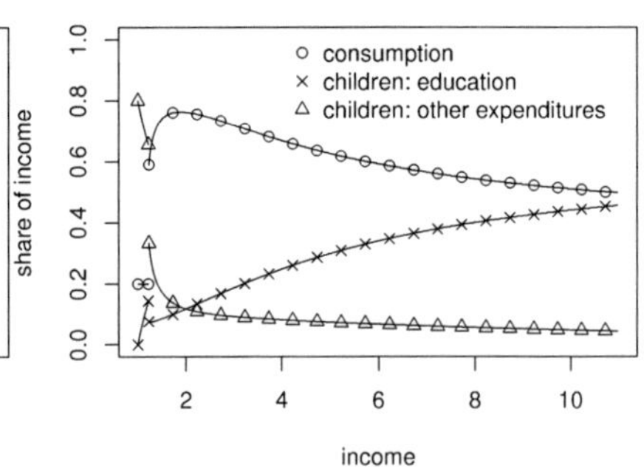

Fig. 4 Shares of consumption, expenditure for education and other expenditure for children on income from the numerical solution

We have done extensive numerical investigations of the problem. The obtained results are consistent with a J-shaped relationship between fertility and level of income. They are presented in Figs. 3 and 4 which have been obtained using the following parametrization: $a_U = 1$, $p_n = p_e = 0.2$, $\alpha_{U\min} = 0.6$ and $\tilde{n} = 4$.

Figure 3 shows how the solution for the number of children changes with the level of income. At low levels of income, the corner solution is that which maximizes utility. When income increases, utility at the internal critical point gradually increases and above a given level of income it becomes a global maximum. The optimal level of fertility jumps from the corner solution to the internal maximum. The figure shows how the qualitative behavior hinted by the analytic investigation holds: increases of income cause first a reduction and then an increase of fertility; the change happens when the marginal utility of consumption becomes sufficiently low.

Figure 4 shows how the types of expenditures considered in the balance sheet constraint (c_t^*, $n_t^* e_t^* p_e$ and $n_t^* p_n$) evolve as ratios of income. At low levels of income there is a phase where consumption has a higher importance than that given to children because an increase of income causes an increase of the share of consumption and a decrease of the number of children. At higher level of income, consumption looses its importance and quantity and quality of children increase.

Till now we have focused on how the single household's behavior changes with income. Next section is dedicated to the aggregate outcome.

4 The Agent-Based Model

In this section we build a model where a collection of households evolves by taking decisions following the rules described in the previous part of the paper. Our aim is to compare the aggregate dynamics of the variables obtained from simulations, especially fertility rates, with the evidence presented in Sect. 2.

To close the model, we need to know how the individual income evolves. Among the available ways to obtain this result, at the present stage of our investigation, we choose to do this in a simple way. We proceed by taking a "partial" analysis point of view, that is by modeling directly the evolution of the income.[3] In doing this a guideline is given by the assumption made in Sect. 3.1 according to which $E[y] = e^{\gamma}$. This requirement have to be fulfilled in a perfect economy with agents having rational expectations. Agent-based modeling is in general at odds with a clock-working view of the economy, however we want our rule to keep track to some extent of the expectation requirement. The assumption on expectation is fulfilled if

$$y_t = E[y_t] \frac{e_t^{\gamma}}{E\left[e_t^{\gamma}\right]}., \tag{7}$$

[3] It is our intention to move towards a "general" analysis in the future. It will be done by introducing in the model other types of agents (firms, workers, researchers) which will interact with each other to produce the aggregate outcome.

that can be easily verified by substituting $E[y]$ for e^{γ} and taking expectation.

As mentioned above, the relationship should be observed in an ideal frictionless economy. To arrive at a dynamic equation that could take into account real world imperfections we use the following rule

$$y_t = \begin{cases} y_{t-1}\frac{e_t^{\gamma}}{E[e_t^{\gamma}]}\varepsilon_t & \text{with probability} \quad \alpha_y \\ E[y_{t-1}]\frac{e_t^{\gamma}}{E[e_t^{\gamma}]}\varepsilon_t & \text{with probability } 1-\alpha_y \end{cases} \tag{8}$$

where ε_t denotes the realization of a random variable. Beside the effect of education, this formulation says the household's income is affected by the previous generation income (y_{t-1}), by externalities represented by the average income of the previous period ($E[y_{t-1}]$) and by a random shock.

The algorithm to simulate the model iterates the following steps:[4]

1. the level of each household's income is computed by using (8),
2. given the level of income, each household "computes" α_U and γ (equation 4),
3. the level of education to be given to each child is computed by using (5),
4. the number of children is computed by numerically searching the maximum of (6) on the set $\{0, 0.5, 1, 1.5, ..., \tilde{n}\}$,[5]
5. the level of consumption is determined by using the balance sheet constraint,
6. the average levels of education and income are computed.

In the remainder of this section we will present and discuss the simulation results. To obtain then we set the initial condition $y_0 = 1$ for all the households, and ε is drawn from a uniform distribution with boundaries 0.995 and 1.005.

Figure 5 shows the dynamics of the average number of children. It could be compared with the total fertility rate displayed in Figs. 1 and 2. In the time series obtained from simulations the features of the real data can be observed: the average number of children smoothly declines for a large time span. It then reaches a minimum and it slightly increases as time goes on. Figures 6 and 7 display the time series of the average income and the average education. These patterns are "typical" of our simulations. To avoid an arbitrary choice of a convenient run, Figs. 5-7 give a summary of one hundred runs. The bold solid black line gives the value of the median, the two gray lines indicate the 20th and 80th percentiles and the two remaining black lines are the minimum and maximum of the hundred values gathered in each simulation time step.

An additional exercise can be done by recording the data of a single run at the individual level. This gives us the opportunity to look at the evolution of the distributions of the households' variables. In particular, the distribution of income and the one of the number of children can be analyzed in the current version of the model.

---

[4] The code in written in Java taking advantage of the facilities provided by the Repast 3 agent-based modeling toolkit.

[5] In the real world the number of children is an integer value. We recall we are working under the assumption that an individual has one single parent, so that the real world variable should be divided by two.

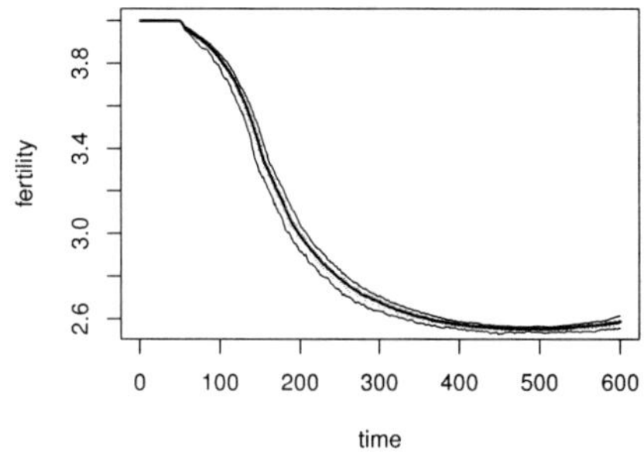

Fig. 5 Average fertility from simulation

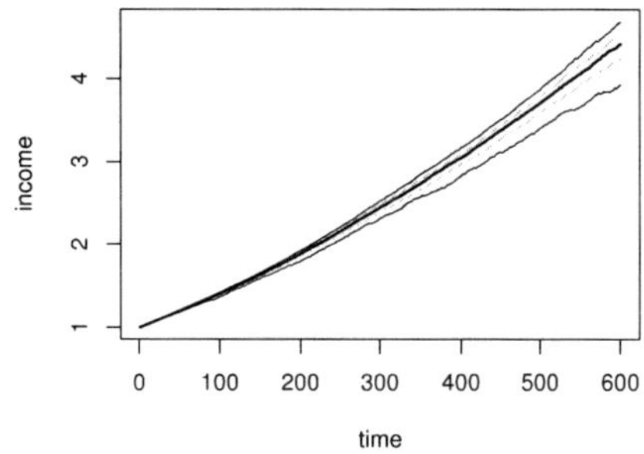

Fig. 6 Average income from simulations

Figure 8 shows how the distribution of income moves to the left and its dispersion gradually increases while time passes (from left to right the three distributions relate to simulation times 100, 200 and 300 respectively). The distributions of the number of children at simulation time 100 and 300 can be seen in Figs. 9 and 10. Households gradually leave the high level of fertility as their income increases and lower level of fertility have higher frequencies as time goes on.

Let us further comment on the income distribution, a topic which has been of interest for economists at least since [15]. The Italian economist observed that the left tail of this distribution (say for $y > y_l$) is characterized by the following survival function:

$$Pr(y > y_l) = \left(\frac{y_l}{y}\right)^a.$$

Preliminary investigations of this topic seem to be in favor of the presence of the Pareto law in the data from simulations (it is perhaps because of the multiplicative random process in equation (8); as pointed out by [**?**] (chapter 14), multiplicative processes can generate fat tailed distributions). A second topic is the relationship between the importance of externalities (the parameter α_y) and the time evolution of the income distribution. A reduction of α_y could be interpreted as stronger policies for reducing income inequalities. Our preliminary results show that a lower level of α_y implies less concentrated distributions, but the growth of the average income slows down. The investigation on the properties of the income distribution and its time evolution in this model is at the moment a work in progress and we avoid presenting the obtained results here. We just inform the reader that this section figures have been obtained by setting $\alpha_y = 0.05$.

5 Conclusions

In recent years a number of economists started to think about a unified model which could generate the different phases of the development process. We believe that

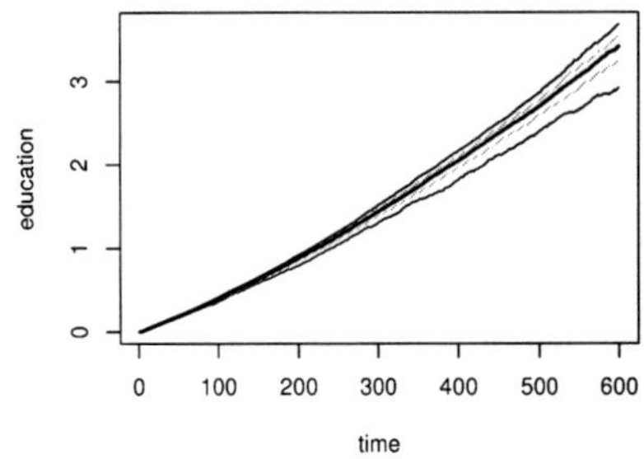

Fig. 7 Average education from simulations

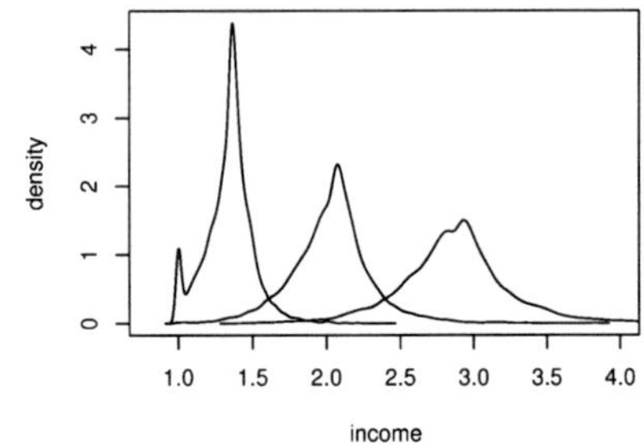

Fig. 8 Evolution of the income distribution

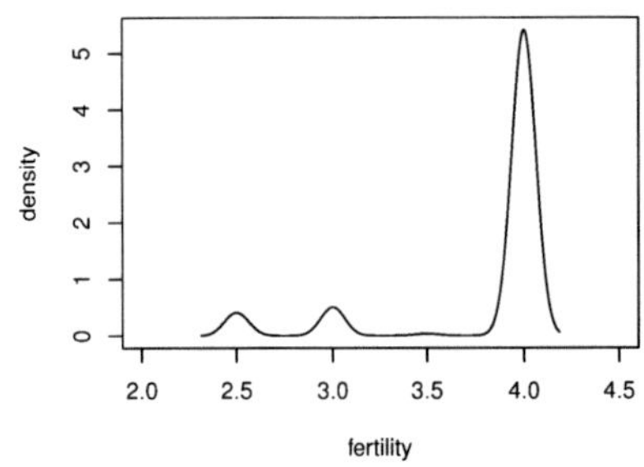

Fig. 9 Distribution of the number of children at time 100

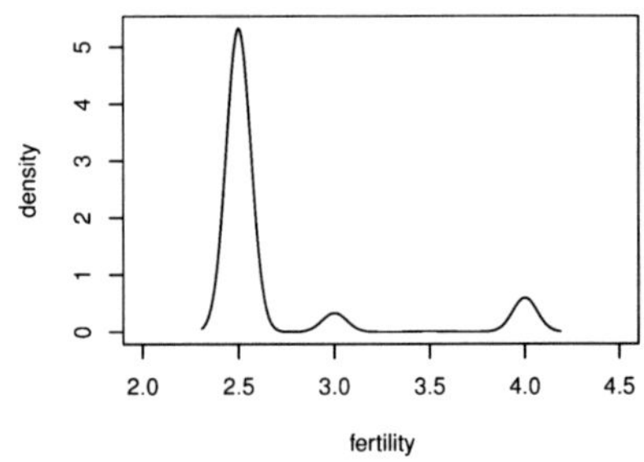

Fig. 10 Distribution of the number of children at time 300

complementing the existing theoretical contributions by using simulation could help in reaching the goal. This paper is a first step to build an agent-based model of the economic development process. We focus on households' behavior by studying in particular the relationship between the available income and the optimal choice on quantity (fertility) and quality (level of education) of children. We first hinted at this choice from a theoretical point of view. Then, with the help of numerical investigations, we have analyzed the individual choice under a more general setting where the characteristics of the household (the parameters of the utility function) change with the level of income. For example, in the model presented above the importance given by the household to her own consumption (represented by the parameter α_U) decreases, while her trust in the education level as determinant of the children's future income (represented by the parameter γ) increases with income. The idea of looking for the determinants of the behavioral parameter has been given by the arguments which can be found in two books. The fist one is that by [8] where the author highlights how the marginal utility of a good is normally increasing (decreasing) at low (high) levels of her consumption. In other terms, the same good looks different to the agent's eyes in different conditions. The second book has been written by [16]. This second contribution can be linked to the previous one because it basically states that the "good" children has the same feature: the social value of children can change (indeed has changed) over time. The results of our numerical investigation give us an opportunity to investigate the relationship between the change of behavior toward consumption and that toward children. In the last part of the paper we build a simple agent based-model and we use it to investigate the aggregate behavior resulting from the evolution of a large collection of households. We show how simulations can replicate the qualitative behavior of the real world total fertility rate dynamics. Furthermore, by collecting data at the individual level from the simulation we are able to analyze how the distribution of key variables like the number of children and income evolves over time.

The potential developments of the model are many. The most important is perhaps the explicit modelization of the supply side of the economy. The level of education is a determinant of the technological progress which in turn affects the efficiency in production and finally employment and the household income. The potentiality of a "general" model could be particularly useful for policy purposes. It could be used, for example, to investigate if the feedback between education and technological improvements creates a situation where multiple equilibria can be observed. The "low" equilibrium represents the poverty (Malthusian) trap while the "high" one is the modern sustained growth regime. Additional efforts can be done to identify the determinants of other parameters of the model (beside α_U and γ) and investigate their role in letting the economy to switch from the "low" to the "high" equilibrium. We refer in particular to the study of how the dynamics of the system change under different assumptions on the time evolution of the cost of growing a child (p_n) and that of education (p_e). It could be the case that a low level of p_e and a high level of p_n are both required for an economy to take off. In other words, beside building schools in poor countries (lowering p_e) authorities should find ways to increase women's wage (increasing p_n) for fighting poverty.

References

1. Barro, R.J., Becker, G.S., (1989) Fertility choice in a model of economic growth. Econometrica **57**(2):481–501
2. Caldwell, J.C. (1997) The global fertility transition: the need for a unifying theory. Population and Development Review **23**(4):803–812
3. Easterlin, R.A. (1978) The economics and sociology of fertility: a synthesis. In: Tilly, C. (ed.) Historical Studies of Changing Fertility. Princeton University Press, Princeton (NJ)
4. Easterlin, R.A. (1980) Birth and fortune: the impact of numbers on personal welfare. Basic Books, New York
5. Galor, O. (2005) From stagnation to growth: Unified growth theory. In: Aghion, P., Durlauf, S. (eds.) Handbook of Economic Growth, volume 1, Part A, chapter 04, Elsevier, 171–293
6. Galor, O. and Weil, D.N. (2000) Population, technology, and growth: From malthusian stagnation to the demographic transition and beyond. American Economic Review **90**(4):804–828
7. Jones, C.I. (2001) Was an industrial revolution inevitable? economic growth over the very long run. Advances in Macroeconomics (The B.E. Press) **1**(02):1–45

8. Karelis, C. (2007) The Persistence of Poverty: Why the Economics of the Well-Off Can't Help the Poor. Yale University Press
9. Kremer, M. (1993) Population growth and technological change: One million b.c. to 1990. Quarterly Journal of Economics **108**(4):681–716
10. Lucas, R.E. (1998) The industrial revolution: past and future. University of Chicago, mimeo
11. Macunovich, D.J. (1998) Fertility and the Easterlin hypothesis: An assessment of the literature. Journal of Population Economics **11**:53–111
12. Macunovich, D.J. (2002) Birth Quake. The baby boom and its aftershocks. The University Press of Chicago, Chicago
13. Mulder, T.J. (2002) Accuracy of the U.S. Census Bureau national population projections and their respective components of change. U.S. Census Bureau, Working Paper Series No. 50
14. Myrskylä, M., Kohler, H.-P., and Billari, F. (2009) Advances in development reverse fertility declines. Nature **460**:741–743
15. Pareto, V. (1897) Cours d'Économie Politique. Macmillan, London
[Sornette, 2006] Sornette, D. (2006) Critical Phenomena in Natural Sciences: Chaos, Fractals, Self-organization and Disorder: Concepts and Tools. Springer.
16. Zelizer, V.A. (1985) Pricing the Priceless Child: The Changing Social Value of Children. Princeton University Press

An Agent-Supported Simulation of Labour and Financial Markets for Migration Processes

Nancy Ruiz, Vicente Botti, Adriana Giret, Vicente Julian, Oscar Alvarado, Victor Perez, and Rosa M. Rodriguez

Abstract The Migration Process is a phenomenon that includes a variety of actors, societies and political issues at different levels. In the migration problem it is then possible to observe complex interactions among different entities: there are links among economics, politics, social, commercial, labour, health, culture, and safety areas. Migration movements may also influence and be influenced by the effects of policies and norms of both sending and receiving countries. One of the key factors that influence the Migration Process behavior is the Labour Market, which is simultaneously affected by Financial Markets. These interactions have been traditionally represented by mathematical approaches that do not allow including flexibility, autonomy, adaptive and pro-activity features that are present into the dynamic and complex real life migration scenarios. On the other hand, the Multiagent System (MAS) paradigm has been successfully applied in studies related to mass movement in complex environments. In this paper a MAS simulation approach is proposed to simulate the migration process and to model micro-level interaction protocols that link Labour and Financial Markets to Migration Processes (MP-LM&FM) in order to observe dynamic behaviours that may emerge at macro level.

Nancy Ruiz, Vicente Botti, Adriana Giret, Vicente Julian , Oscar Alvarado
Universidad Politecnica de Valencia, Camino de Vera s/n, 46020,Valencia, Spain
e-mail: {nruiz,vbotti,agiret,vinglada,oalvarado}@dsic.upv.es

Victor Perez, Rosa M. Rodriguez
Boeing Research & Technology Europe, Cañada Real de las Merinas, 1-3 Building 4 - 4th floor 28042 Madrid, Spain
e-mail: {victor.perezvillar,rosamaria.rodriguez}@boeing.com

M. LiCalzi et al. (eds.), *Progress in Artificial Economics*, Lecture Notes in Economics and Mathematical Systems 645, DOI 10.1007/978-3-642-13947-5_20,

1 Introduction

Migration is a phenomenon that may influence policies and norms of both sending and receiving countries. Furthermore, Labour and Financial Markets behaviors are also affected by migration. Thus, the study of migration processes and their interactions with Labour Markets, the influence of the Financial Markets (MP-LM&FM) at micro-level, and the global behaviour that emerges at macro-level are key issues in managing migration problems.

The Migration Process includes a variety of actors, societies and political issues at different levels. There are links among economics, politics, social, commercial, labour, health, culture, and safety areas. Then, in the migration problem it is possible to observe complex interactions among different entities. The behaviour patterns can be defined as complex because there are n dynamic variables, where the participants are people with autonomous and intelligent behavior, and the environment constantly evolves. Therefore, as some researchers point out [20, 10], one of the key factors that influences the Migration Process behaviour is the Labour Market. Moreover, Labour Markets are simultaneously affected by Financial Markets. However state of art research studies on the migration problems and their relation with the Labour and Financial Markets are based on mathematical models [2, 19, 23] in which it is difficult to include flexibility, autonomy, adaptive, heterogeneity, self-interested behaviors and pro-activity features that are present into the dynamic and complex real life migration scenarios.

The Multiagent paradigm allows to define adaptive, autonomous, interdependent, networked entities like "social agents". A social agent acts according to inputs, decision rules and protocols in response to the influence of its environment. The agent population dynamics depend on the network properties like clustering, hubs, and patterns. A social agent can influence his neighbors in response to the local influence that it receives. Then, Multiagent theory allows to represent actions and interactions (as behaviour patterns) of autonomous entities in an environment. In a Multiagent System (MAS) the global behaviour emerges from the interactions and the properties of the agents. Thus, a MAS model allows to simulate simultaneous behavior of multiple agents to show and predict actions of complex phenomena. The Agent-based Social Simulation (ABSS) studies the social phenomenon by using computational models. These models may represent people and their interactions as agents [9]. The ABSS is focused on the emergence properties of large agent groups that react to its environment following a set of rules.The MAS paradigm has been applied in studies related to mass movement by defining agent-based models to simulate the rural-urban movement according to social learning [18, 7]. Many models have been developed in this field, such as the Ethnic Migration Model (EMM) [13] and the Schelling Segregation Model (SSM) that shows the emergence of socio-spatial patterns [17]. Then, it is possible to observe that MAS approaches allow to represent the behavior of the interactions and influence among complex environments like the Migration Process, the Labour and Financial Markets at micro-level and observe the global behavior that may emerge at macro-level.

As a result of a study related to the migration process and their relation with the Labour and Financial Markets from the point of view of MAS, we have identified behaviour patterns that can be represented using agent-based models. This approach allows to use predictive techniques to represent the complex environment of links among the Migration Process, Labour Market and Financial Market models. These processes have been translated into simulation models that can be used to predict and react to emergent situations. In this paper a MAS simulation approach is proposed to simulate the migration process and to model micro-level interaction protocols that link Labour Market, Financial Market, and migration processes (MP-LM&FM) in order to observe dynamic behaviours that may emerge at macro level.

In this paper representative models of MP-LM&FM influence are presented. Therefore, a proposal of an agent-supported model of MP-LM&FM influence into an agent-based architecture is presented. Finally, the features of an agent-based simulation environment that was developed as a technological Demonstrator to validate the agent-supported architecture are shown.

2 Labour and Financial Market Models

The Labour Market behavior according to Neoclassical Economic Theory [14] (where migration occurs as a response to regional differences in income opportunities, such as job opportunities) is a key factor that influence the Migration Process behavior. The Labour Market behavior can modify not only the target country but also the migration routes. People make decisions based on different factors such as distance, price, risk, and mainly the labour requirements. When a Labour Market requires a large number of people and offers “good” payment in a specific place, people will try to go to that place instead of another. At the same time, Labour Market is linked to the Financial Market behavior. The external supplies of an Industry take into account the required inputs of other industries or sectors and, at the same time, the internal supplies consider the labour resource requirements (inter-industry flows). Thus, the labour resources required by the Industry are provided by the Labour Markets and Financial Market is composed by Industries that influence the Labour Market behaviour. Therefore, any change (increment or decrement in demand or production) in the Financial Market behavior will be reflected at the Labour Market level. The behavior of Migration Process, Labour Market and Financial Market represents the system behaviour at micro-level and from their interactions the system behaviour at macro-level emerges (see Sect. 3 and Fig. 2). In this work, Labour Market is graphically depicted as a layer below the Financial Market because of its dependence on the Financial Market behaviour.

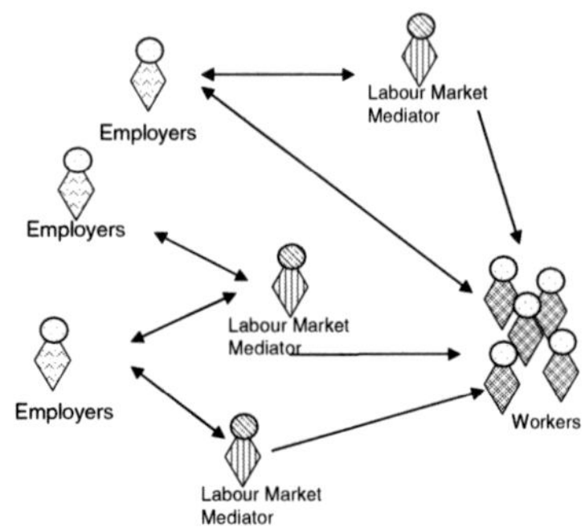

Fig. 1 Labour Market entities

2.1 Labour Market

Labour economics seeks to understand the functioning and dynamics of the market for labour. Labour markets run through the interaction of workers and Labour Resource Providers (Employers) [5, 3]. Labour economics observes the suppliers (workers) and the demanders (employers) of labour services and attempts to understand the resulting pattern of wages, employment, and incomes. In this section a representative model of a Labour Market is presented. Human capital is created and put into use in Labour markets. The structure of the Labour market is therefore critical for the quantity and quality of human capital that is created and for the uses to which it is put. Under dynamic efficiency of the Labour market the emphasis is shifted to the efficiency of the Labour market in the provision of adequate resources for education and training and to the recruitment of enough people for these purposes. According to Todaro and Harris [20, 10] the probability of eventually finding a job in a specific sector is a key factor that influences the migration to specific places (Industries) or sectors [22]. Thus, the Labour Market requires people with specific skills to meet their needs and thus the Labour Market requirements influence people when they want to migrate during their making decisions process.

A Labour Market model (Fig. 1) includes at least two main entities: Employers and Workers. Labour demand is a derived demand; in other words the employer's cost of production is the wage at which the business (or firm) benefits from an increased output or revenue. The process of allocation of vacancies to fill the labour demand usually is done by Employers that can request help to Mediators to get their labour resources. Mediators are specialized people that search and select the better labour resources according to the employer requirements (skills, payment, age, gender, etc.).

2.2 Financial Market

Computable General Equilibrium (CGE) models are probably the most utilized tool for development planning and macro policy analysis [15]. Bandara [1] and Jorgenson [11] both recognize that the beginnings of CGE (or 'empirically estimated economy-wide') modeling started with the work of Leontief in the 1930's. The Leontief Model has been selected due to the fact that it is a reference model in economics research environment. It is worth to highlight its flexibility, simplicity and the successful results obtained in economic and social studies. In this section a brief description of a Financial Market based on Leontief Model is presented. Leontief created an accounting system that encompassed "all branches of industry, agriculture, and transportation (and) also the individual budgets of all private persons" [12]. Leontief set up his own structure for a 'national account' of the economy wherein all sectors produced goods or services that were fully consumed by another sector. Input-output Analysis was partly inspired by the Marxian and Walrasian analysis of general equilibrium [21] via interindustry flows. The goal of the model is to allow the Economists to predict future production level of each industry and satisfy future demand of a variety of products.

3 Agent-Supported Simulation Architecture

Based on the models and interactions previously described (Sect. 2), this section presents an agent-supported simulation architecture that represents the MP-LM&FM influence. The agent-supported architecture translates all the actors of the Migration Process, Labour Market and Financial Market into Agents to represent their micro-level behaviour. Thus, from the interactions among them the macro-level behavior of the global system will emerge (Fig. 2). At micro level, the migration process is directly influenced by the Labour Market behavior. The Labour Market manage the available vacancies that can be filled by migrants. Labour Markets interact with Financial Markets due to the available vacancies offered per Labour Market. Thus at macro level, it is possible to observe the global behavior of the system as a result of the agent interactions at micro-level.

The Agent-supported Simulation Architecture simulates the Migration Process, Labour and Financial Markets like a complex social system with dynamic behaviour patterns of their participants in specific regions. To this end, the region includes specific places called "nodes" in which it is possible to observe migration activities like departure/arrival of Migrants, Labour Market Management, Security Patrols, and Financial Market Management. The simulation includes predefined routes that Migrants can use. The routes connect specific region nodes. Thus, the architecture includes communication protocols that the agents use according to specific scenarios. The main scenarios that can be simulated include: a) Search of transportation to move along migration routes done by Migrants, b) Negotiation among Transportation Service Providers and Migrants, c) Movement of migrants with different

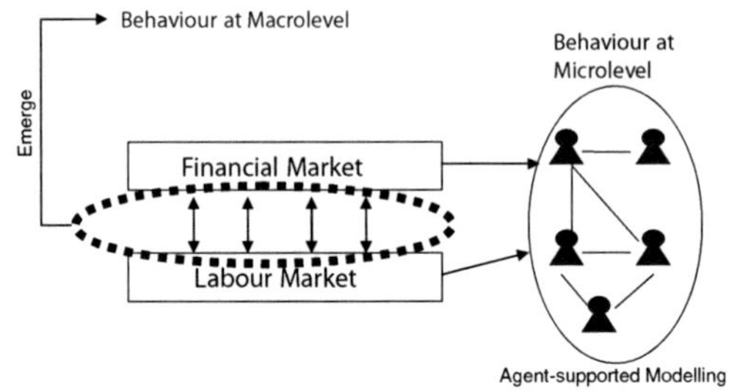

Fig. 2 Agent-representation of behaviour at micro and macro level

transportation mechanism, d) Search a job to earn money done by Migrants, e) Execution of negotiations processes to link position vacancies to workers managed by Labour Market Mediators, f) Patrol of arrived transports at specific places done by Security Forces, g) Capture of migrants that arrive to a specific place done by Security Forces, h) Verification of documentation of captured Migrants done by Security Forces, i) Control of the required inputs (supplies) of the Industries for their production according to their demand done by Industry Agents, j) Supply Labour Resources to Industries through Labour Resources Providers and Labour Market Mediators, and k) Supervision and management (done by the manager of financial markets) of changes in inputs and outputs requirements, demand and production of all industry agents.

The Agent-supported Simulation Architecture also allows the user to track the behaviour of agents in specific nodes by storing data that show arriving/departing migrants, position vacancies and dynamic payment by capabilities, results from the security patrol execution (migrants that have been captured and released), and management of the Financial Market. Moreover, the Agent-supported Simulation Architecture allows the generation of simulation data that can be post-processed with external tools. This data can be used to analyze previous situations and predict future actions. The main issue of this architecture is therefore the integration of Labour and Financial Market into the migration process. Next sections detail its main features.

3.1 Agent-Supported Labour Market

In the Agent-supported Labour Market it has been defined three types of roles: Labour Resource Providers, Workers, and Labour Market Mediators.

Labour Resources Provider Agent. Labour Resource Providers offer a number of vacancies that require a specific skill and a minimum grade of skill, gender and age. Each vacancy offers a given payment per hour and a defined contract duration. Then payment is the maximum amount that Labour Resource Providers offer.

Labour Market Mediator Agent. A Labour Market Mediator manages the allocation of vacancies offered by the Labour Resource Providers according to the vacancy requirements and the available Workers. Labour Market Mediators use negotiations mechanisms to link vacancies to workers. Thus, Labour Market Mediator uses as inputs: a) Vacancies offered by Labour Resource Providers, and b) Worker Requests. These inputs are stored and managed during allocation process. The Labour Market Mediator manages outputs related to links between offered and requested vacancies and informs Labour Resource Providers and Workers about these links.

Worker (Immigrant) Agent. Workers are looking for vacancies according to their presumed skills and a minimum requested payment.

The vacancy allocation process is the key activity in the Labour Market. It is guided by negotiation processes among the Labour Resource Providers and Workers and is controlled by the Labour Market Mediators. The multiagent paradigm makes possible to use any negotiation protocol for automating negotiation in an easy way (i.e. double-auction mechanism [8]). These protocols define the rules that all participants must follow during the vacancy allocation process. In this paper, the Labour Market definition is focused on specifying a negotiation mechanism during the vacancy allocation process. This mechanism can vary according to the features of the node (i.e. management of regular or irregular job vacancies, fixed or dynamic salary, etc.).

3.2 Agent-Supported Financial Market

This section presents the Financial Market Model presented in 2.2 (Leontief Model [12]) using the agent paradigm. Industries of Financial Market require to control inputs and outputs to cover demand and production. The industries are represented like agents that request inputs to other agents. In the Financial Market relations among industries are represented like complex interactions thanks to the agent paradigm. The Agent-based Financial Market is composed of three type of actors:

Industry Agent. This Agent specifies the coefficient inputs required according to the input-output matrix of Leontief Model. The requirements are linked to the production and demand levels. Thus, the Industry Agents supervise if the inputs required are affected by changes on production according to the current demand. When a change occurs, this agent will modify the required inputs according to its coefficient inputs (Leontief Model). The Industry Agent informs both the Labour Resources Provider Agent about the change of Labour requirements and the Manager of Financial Market which adjusts the input-output matrix.

Labour Market Mediator Agent. This Agent searches and supervises that the Labour requirements requested by the Industry Agent are fulfilled. Then, this agent interacts with the Labour Market. The Labour Market Mediator Agent requests a number of vacancies of a specific skill and expertise grade, duration and

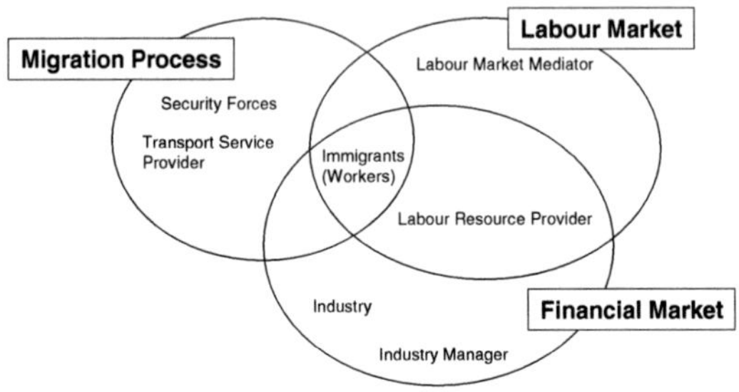

Fig. 3 Agent-interaction among Migration Process, Labour and Financial Markets

payment per hour (acting like a Labour Resource Provider). If Industry Agent informs on a change of Labour requirements, the Labour Market Mediator Agent updates the new requirements. Labour Market Mediator Agents participate at Labour Markets and Financial Markets doing different activities.

Manager of Financial Market. This Agent supervises the Industry Agents behavior. If some Industry Agent changes its Demand or Production, the Manager of Financial Market informs the Industry Agents about the changes of input requirements according to the Matrix inputs defined in the Leontief Model.

Due to the continuous influence between the Agent-supported Labour Market and the Agent-supported Financial Market (Section 2), the internal behavior of the Financial Market triggers a reaction in the internal behavior of the Labour Market (Fig. 3). At the same time, changes at Labour Market trigger a reaction at the Migration Process.

4 ARGOS: An Agent-based Demonstrator

An agent-based Demonstrator called ARGOS has been developed to simulate and validate the Agent-supported Architecture presented in Section 3. The simulation validates the scenarios included into the architecture by defining an initial state of the migration process. Thus, the migration problem has been bounded in a specific region. The initial state includes: a) specific routes to connect places, b) a given number of immigrants with specific profiles that allow to identify clusters of immigrants, c) Labour Markets that influence the preference of immigrants to select specific migration routes in specific places, d) the negotiations processes (to link position vacancies to workers) based on auction mechanisms [6] (the double-blind auction), and e) the use of strictness level that Security Forces apply when supervising the documentation of captured immigrants. ARGOS has a web interface that includes three modules (Fig. 4): a) Argos Control (a module that allows controlling the simulation execution of the behaviour of the agents) , b) Argos Player (a module that allows playing current and previous simulations with a flexible and easy-to-use interface), and c) Argos Data (a module that allows analyzing the generated data

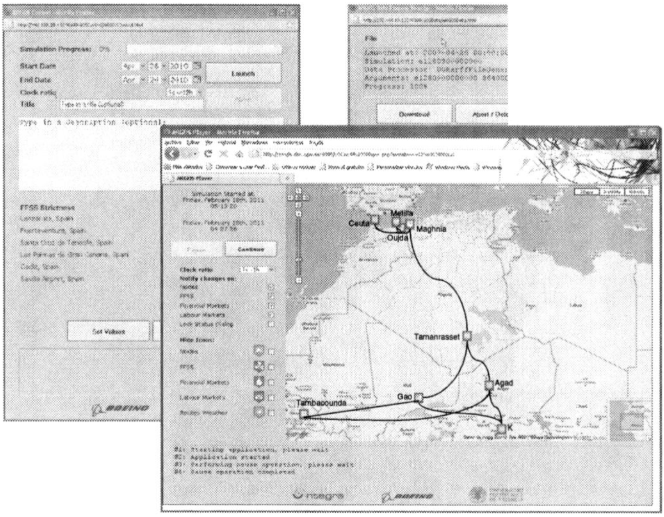

Fig. 4 ARGOS Control - Argos Player - Argos Data

during simulation executions). ARGOS also allows the User to track the behaviour of agents in specific nodes by using graphic data that show among others arriving/departing Migrants, position vacancies and dynamic payment by capabilities, changes on Financial Markets and their influence translated as available vacancies that manages specific Labour Markets and results from the security patrol at nodes. Argos Control also generates simulation data of the behavior Migration Process, and Labour and Financial Markets for external analysis. This generated data can be used to analyze previous situations and to predict future actions. Then, Argos Data allow the analysis of the historical data generated by Argos Control.

Argos Player displays the behaviour of Migration Process and the influence of the Labour and Financial Markets using the data generated by the Argos Control module (Fig. 4). Argos Player allows to observe how changes of Financial Market Demand in a node (that appears at specific time) generate requests of human resources to cover its production and demand. The required human resources are requested with specific skills to the Labour Market. In this way, it is possible to observe how the level of immigrants in that node increases due to the availability of vacancies.

5 Experimentation

ARGOS has been used to simulate the migration process in a specific region where the main source of migrants is Senegal and Nigeria and the main goal of these migrants is to reach Spain. The region includes specific places called "nodes" in which it is possible to observe migration activities like departure/arrival of Migrants, Labour and Financial Markets Management, and Security Patrols: Tambacounda

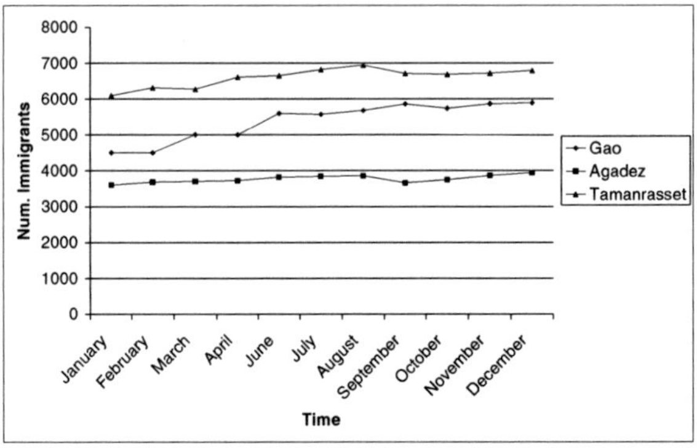

Fig. 5 Migration behaviour according to Labour Markets

(Senegal), Gao (Mali), Kano (Nigeria), Agadez (Niger), Tamanrasset (Argelia), Maghnia (Algeria), Oujda (Morroco), Melilla and Ceuta (Spain). These nodes are connected by migratory routes see Fig. 4. There are Labour Markets at Gao, Agadez, Tamanrasset and Oujda. There are Security Forces at Maghnia, Oujda, Ceuta and Melilla. Financial Markets are located in Oujda. The scenario shows one of the migration routes chosen in the late 90's by sub-saharan Africans to reach Europe. Before reaching destination, many migrants remain in transit countries for months or even years to earn money for their own subsistence, and for financing the rest of the trip to Europe. ARGOS also shows climate changes at the region. Migrants takes into account the climate when they select a transport. The Labour and Financial Markets has been limited to five representative economic sectors: agriculture, lifestock farming and fishing, energy, industry, building, and services. During simulation, Labour and Financial Markets at nodes are composed by different economic sectors [16]. Agriculture, labour lifestock farming and fishing sector has a demand calendar that is linked to the product type (i.e. orange, grape, tuna, etc.). Financial Markets manages Input-output matrixes that include the active economic sectors at each node. Thus, for the simulated scenario, the Labour Markets are composed as follows: in Agadez there are agriculture, and industry sectors, in Gao and Tamanrasset there are agriculture, industry, and services sectors. Moreover, the Financial Market at Morroco has agriculture, lifestock farming and fishing, energy, industry, building, and services as the more representative economical sectors. The Financial Markets used are based on the Leontief Model [4].

During simulation execution Migrants query the Labour Market conditions at different nodes, when they search the best route to get money and reach to their final node (i.e. cities in Spain). Migrants select the best migration route taking into account some data: price, duration, risk of the route (linked to geographical features), the successful probability to reach the target node related with climate data and their knowledge about the target node. Two important features drive the migrants' choice about the node to be reached: the existence of a labour market and the presence of job opportunities that match their skills. In order to evaluate the Labour and

Financial Markets influences over the migration routes we have conducted 100 executions of the simulation scenario described above. These executions simulate the migration behaviour from January 1999 to December 1999. Figure 5 shows the average number of Migrants that move through the defined routes during that year. It is also possible to observe that in Tamanrasset and Gao during summer time, when their sector of services increase their labour resources demand migrants select more migration routes that reach to these nodes than Agadez, which does not have service sector. This behaviour allows to validate that the migrants preference those routes in which nodes offer more job opportunities.

6 Conclusions

In this paper a MAS simulation approach has been presented, the proposed approach allows to model micro-level interaction protocols in order to simulate migration processes. The main focus of this paper is the integration of a labour and a financial markets into an Agent Based Migration architecture. This integration allows to simulate the influences of these two markets into migration processes. Finally, the features of an agent-based simulation environment called ARGOS which allows the validation of the agent-supported architecture has been described. The simulation approach has been successfully used as a tool for migration model analysis. Furthermore, the models of migration processes, Labour and Financial Markets have been translated into simulation models that are used to predict and react to emergent situations.

Agent Technology has been used to support the decisions process during migration management. In order to validate this analysis, it has been included the Labour and Financial Markets behaviour into the agent-based simulation environment previously developed. ARGOS has been used to validate the use of the multiagent paradigm to represent the complex social interactions at micro-level and observe the global behavior that emerge at macro-level as a result of the interactions among the MP-LM&FM. The simulation environment called ARGOS includes the following key entities of the migration problem: Migrants, Transport Service Providers, Security Forces, Labour Market Mediators, Employer Services Providers, and Financial Market Managers. These entities are implemented by means of agents with complex behaviour. Several tests have been made using ARGOS in order to analyze the behaviour of the Labour and Financial Markets (MP-LM&FM) and their influence over the migration process. Thus, it is possible to observe the influence of Labour and Financial Markets over changes on migration routes. Some future work includes: the definition of more specific profiles of migrants, increase the migrant knowledge of the environment in order to takes into account more attributes to select the best migration routes, add more nodes with labour and financial markets, and define metrics to manage the influence at micro and macro level.

Acknowledgements This research is partially funded by the Spanish Center for the Development of Industrial Technology (CDTI) as part of the 2010 Ingenio Program. This support is gratefully acknowledged.

References

1. Bandara, J.S. (2001) Computable General Equilibrium Models for development policy analysis in LDCs. In: Journal of Economic Surveys **5**(1) March, 3–69
2. Borjas, G.J. (1993) The Impact of Immigrants on Employment Opportunities of Natives. In: BOECD The Changing Course of International Migration. Paris: OECD, 191–198
3. Borjas, G.J. (2005) Labor Economics. McGraw-Hill
4. Bussolo, M., Roland-Holst, D. (1993) A detailed Input-Output table for Morroco, 1990. Working Paper No.90, OCDE/GD(93)180
5. Cahuc, P., Zylberberg, A. (2004) Labor Economics, MIT Press
6. Cliff, D. (1998) Evolving Parameter Sets for Adaptive Trading Agents in Continuous Double-Auction Markets. In: Proc.Workshop Artificial Soc. and Computational Markets
7. Espindola, A.L., Silveira, J.J., Penna, T. (2005) A harris-todaro agent-based model to rural-urban migration. Brazilian Journal of Physics **(36-3A)**: 603–609
8. Friedman, D., Rust J. (1992) The Double Auction Market: Institutions, Theories and Evidence. Addison Wesley
9. Gotts, N., Polhill, J., Law, A. (2003) Agent-based simulation in the study of social dilemmas. Artificial Intelligence Review **19**:3–92
10. Harris, J.R., Todaro,M.P. (1970) Migration,Unemployment and Development: a Two-Sector Analysis.American Economic Review **60-1**:126–142
11. Jorgenson, D.W. (1984) Econometric methods for applied general equilibrium analysis. In: Applied General Equilibrium analysis. Cambridge,UK: Cambridge Univ. Press
12. Leontief, W. (1951) The Structure of the American Economy, 2d ed., White Plains, 1919–1939. NY, International Arts and Sciences Press
13. Makowsky, M., Tavares, J., Makany, T., Meier, P. (2006) An agent-based model of crisis-driven migration. In: Proc.Complex Systems Summer School. Santa Fe Institute, New Mexico
14. Menger, C. (1976) Principles of Economics. Ludwig von Mises Institute, USA
15. Mitra-Kahn B (2008) Debunking the Myths of Computable General Equilibrium Models (SCEPA Working Paper 2008-1). Schwartz Center for Economic Analysis. London
16. National Statistical Institute from Spain. `http://www.ine.es`
17. Schelling, T. (1969): Models of segregation. American Economic Review (59-2):488–493
18. Silveira, J.J., E.A., Penna, T. (2006) Agent-based model to rural-urban migration analysis. Physica A: Statistical Mechanics and its Applications **(364)**: 445–456
19. Simon, J.L. (1989) Basil Blackwell. Oxford
20. Todaro, M.P. (1969) A Model of Labour Migration and Urban Unemployment in Less Developed Countries. American Economic Review **59-1**: 138–148
21. Walras, L. (1958) Elements of Pure Economics. Routledge Library Editions, London
22. White, M. J., and Lindstrom,D.P. (2005) Internal migration. In: Handbook of population. Springer, 347–382
23. Zimmermann, K.F. (1993) Unemployment and Migration. In: Europe and Global Economic Interdependence. Bruges: European Interuniversity, 25–52

Sensitivity Analysis of an Agent-Based Model of Culture's Consequences for Trade

Saskia L.G.E. Burgers, Gert Jan Hofstede, Catholijn M. Jonker, and Tim Verwaart

Abstract This paper describes the analysis of an agent-based model's sensitivity to changes in parameters that describe the agents' cultural background, relational parameters, and parameters of the decision functions. As agent-based models may be very sensitive to small changes in parameter values, it is of the essence to know for which changes the model is most sensitive. A long-standing metamodeling-based approach of sensitivity analysis is applied to the agent-based model. The analysis is differentiated for homogeneous and heterogeneous agent populations. Intrinsic stochastic effects of the agent-based model are taken into account. The paper describes how an appropriate regression model has been selected and analyses the parameter's variance contributions in general and in specific cultural settings.

1 Introduction

Agent-based models are known to be very sensitive to parameter changes in some ranges of the parameter space. Small changes in parameter values may have dramatic consequences for the state of the system, while changes in other parts of the

Saskia L.G.E. Burgers
Biometris Wageningen UR, Postbus 100, 6700 AC Wageningen
e-mail: saskia.burgers@wur.nl

Gert Jan Hofstede
Wageningen University, Postbus 8130, 6700 EW Wageningen
e-mail: gertjan.hofstede@wur.nl

Catholijn M. Jonker
Delft University of Technology, Mekelweg 4, 2628 CD Delft, NL
e-mail: c.m.jonker@tudelft.nl

Tim Verwaart
LEI Wageningen UR, Postbus 29703, 2502 LS den Haag, NL
e-mail: tim.verwaart@wur.nl

M. LiCalzi et al. (eds.), *Progress in Artificial Economics*, Lecture Notes in Economics and Mathematical Systems 645, DOI 10.1007/978-3-642-13947-5_21,

parameter space have little effect. This property of multi-agent systems is usually referred to as non-linearity. It is not just a property of agent-based models. It is a general property of complex systems such as ecosystems, climate, and the economy. Non-linearity may lead to abrupt changes in the state of systems, and this property invites to the application of agent-based models to simulate non-linear effects such as catastrophic events in evolution or economics [3, 12]. We may conclude that non-linearity is not a bad property of agent-based models. It is a general property of complex systems that complicates the work of modelers of such system.

In general it is considered good modeling practice to perform sensitivity analysis as a part of model verification [16]. In the case of agent-based models, two reasons urge to perform extensive sensitivity analysis: great uncertainty about actual values of model parameters and non-linearity. For instance, in the model discussed by Kirman [11], a tipping point between loyalty to trade partners and shopping behavior exists, depending on the value of the loyalty parameter β. If one wants to apply such a model in multi-agent models of markets, the agents have to be configured with actual values for β. For some range of low values of β, the value will not have an effect on the shopping behavior of a single agent. Around some critical value of β, there is an abrupt change, and there is a relatively small range of increasing loyalty. For a large range of higher values for β, the behavior is invariably loyal. As a result, depending on the actual distribution of β in the agent population, the efficiency of an artificial market may be very sensitive to small changes in the distribution of β, or may be rather insensitive to even larger changes. However, it is hard to predict the actual distribution of β for a particular context.

Because of the combination of non-linearity and uncertainty about parameter value distributions, extensive sensitivity analysis is a sine qua none for research with agent-based models. Before a conclusion can be drawn on the basis of an agent-based model, the modeler must search for the regions in parameter space where stable, maybe inactive, states of the system occur and where the model is insensitive to parameter changes, regions where tipping points occur and system behavior changes dramatically in case of small parameter changes, and regions where the system is more or less proportionally sensitive to parameter changes.

This paper presents the approach and results of extensive sensitivity analyses of a model of culture's effects on international trade. The multi-agent model is based on a model of a trade game that allows for experimental data collection on trust in supply chains with asymmetric quality information [9]. The model is based on transaction cost economics [19]. The agents' activities cover partner search, negotiation, and, if negotiation leads to a contract, truthful delivery or opportunism, taking advantage of the information asymmetry. Their counterparts may either trust the deliveries, or incur cost to monitor and enforce contract fulfillment. The agent model of Jonker et al. [9] has been refined and extended with differentiation of agent behavior according to cultural background [5, 6, 7]. For this purpose, rules were formulated for adaptation of default model parameters based on Hofstede's five dimensions of culture [4].

Sensitivity analysis is performed on the extended model. A systematic sensitivity analysis can serve several purposes: improve the understanding and reliability of model results; reveal effects of parameter variations; guide simplification and refinement of the model [15]. This paper focuses on the effects of parameter variation. The following are the main questions for the sensitivity analysis.

1. Which areas in parameter space result in realistic behavior?
2. Which parameters have significant effects for which outputs?
3. Which interactions between culture and other parameters are important?
4. Are the answers different between aggregate and individual level?

Sensitivity analysis basically consists of a statistical analysis of the effect of input variations on model outputs. Richiardi et al. [15] identify types of variations of inputs. These types can be grouped into (I) variations of random seed and noise level, (II) variations of parameter values, (III) variations of the model, e.g. agent's decision functions, data aggregation, time scale and sample size. The present paper focuses on the first two groups of variation. It studies the effect of intrinsic variation caused by the stochastic nature of the model and the effect of external variation of model parameters and of culture. The sensitivity analysis approach is based on Jansen et al. [8] and Saltelli et al. [17], applying two principles:

1. meta modeling of results of parameter sets drawn at random from the joint distribution;
2. analysis of contributions of Top Marginal Variance (TMV) and Bottom Marginal Variance (BMV) of individual parameters or groups of parameters to the variance explained by the meta model.

Section 2 of this paper introduces the model and the parameters taken into account. Section 3 presents the approach of sensitivity analysis and discusses the special issues with respect to unit of analysis (system vs. individual) and heterogeneity. Section 4 presents results for some observable statistics at system and individual level. Section 5 concludes the paper with an evaluation the applied method.

2 Trading Agents with Cultural Background

The model analyzed in this paper simulates trading agents operating in a game [9]. The agents may trade with each other, are free to select or refuse a partner, negotiate or quit negotiation if they do not expect a satisfactory conclusion, and, in case of successful negotiation, exchange a commodity. The special thing about the game is that commodities have high or low quality and that the seller is informed about the quality, which is invisible for the buyer. A buyer can either trust a delivery or (at the cost of a fee) offer it to the tracing agency that reveals the real quality and in case of deceit punishes the deceiver by a fine. Another option for the buyer is to have the seller trace the commodity in advance and add the tracing report as a quality certificate. The tracing fee for sellers is lower than it is for buyers. The strategies a buyer can chose are: (1) buy low quality (no risk), (2) trust, (3) require certification, (4) trace random samples, or, (5) in addition to random tracing, negotiate that some refund will be made in case quality turns out to be non-compliant.

Details of the models of the agents' activities and the effects of culture have been described in earlier papers [5, 6, 7]. For each of these activities, a model of the agents' decisions is selected from social sciences or artificial intelligence literature. For instance, for partner selection, the model of Weisbuch et al. [18] is used; for negotiation Jonker and Treur's ABMP architecture [10] is selected. The decision models' parameters included in the sensitivity analysis are listed in Table 1.

Table 1 Trading agent's activities and model parameters and variables that are adjusted according to an agents' cultural background [5, 6, 7]; the table also specifies the value range considered in the sensitivity analysis

Activity	Parameter or variable	Value range
Partner selection	Loyalty	0.5...1.5
	Learning	0.001...0.999
	Preference (initial value)	0.001...0.999
Negotiation	Concession factor	0.001...0.999
	Negotiation speed	0.001...0.999
	Impatience	0.001...0.999
	Quality preference	0.001...0.2
	Risk aversion	0.001...0.2
Deceit and trust	Minimal honesty	0.001...0.999
	Honesty decay factor	0.001...0.999
	Trust (initial value)	0.001...0.999
Belief update	Negative update factor	0.001...0.999
	Endowment factor	0.001...0.999

In the agent model the decision functions are influenced by a set of rules that take Hofstede's cultural dimensions and some culturally relevant relational characteristics into account [5, 6, 7]. The indices of the cultural dimensions are:

- PDI (power distance);
- UAI (uncertainty avoidance);
- IDV (individualism);
- MAS (masculinity);
- LTO (long-term orientation).

The relational characteristics taken into account are group distance (i.e. absence of common group membership) and societal status of the agent and of its partner. Cultural indices and relational characteristics are represented as real values in the range [0...1]. For the sensitivity analysis they are drawn from the range [0.001...0.999].

3 Sensitivity Analysis Approach

The sensitivity analysis reported in this paper is regression-based: a meta model in terms of the input parameters is fitted to an output variable. The output is produced by simulation runs using input parameter sets generated by Monte Carlo sampling. Monte Carlo sampling of the parameter sets aims to cover the range of all parameters efficiently and to avoid multicollinearity.

The relative importance of individual input parameters on output variables is assessed by decomposition of the variance of the output variable. The key issue in this approach is to find a regression model that can serve as a basis for decomposition of variance. Any type of regression may be applied, e.g. linear regression including polynomial and interaction terms [14] or regression with smoothing splines [2] as a form of nonparametric regression, as long as it explains a great deal (preferably at least 90%) of the output variance.

Jansen et al. [8] define the top marginal variance (TMV) of an input as the variance reduction that would occur if the input would become fully known. The bottom marginal variance (BMV) is the variance that the meta model can not explain without the input parameter. TMV and BMV of an input variable are equal if and only if that variable is not correlated with any other variable. Comparison of TMV and BMV can be used to check for multicollinearity unless interaction-terms are important. If interaction-terms are taken into account in the regression model, the BMV is defined as the variance that cannot be explained without the input parameter and all interaction terms including this parameter.

In this sensitivity analysis three sources of variance are studied:

1. cultural and relational factors, used to adapt the decision making to culture,
2. the default values of the parameters mentioned in Table 1,
3. stochastic effects caused by variation of random seed.

The approach proposed by Jansen et al. [8] was developed for equation-based models, in which there is a single level of aggregation. When analyzing multi-agent systems, the unit of analysis has to be decided: system performance at aggregate level or individual agent performance. The present study observes outputs at aggregated level for simulations with homogeneous agent populations and at individual agent level for simulations with heterogeneous populations.

Data generation proceeds as follows. The first step is to draw input parameters sets from the joint distribution of all model parameters. As the goal is to study the effects of parameter variation and there is no accurate information on actual parameter distributions, we draw values at random from uncorrelated uniform distributions, ranging as indicated in Sect. 2. The resulting parameters sets are used to initialize trading agents for simulation runs. In order to analyze intrinsic stochastic effects, model runs are repeated with equal parameter sets but different random seed.

The following outputs are observed:

- number of transactions;
- number of failed negotiations;
- average duration (number of rounds per negotiation)
- number of high quality transactions;
- number of deceitful transactions;
- number of traces requested;
- number of fines issued by the tracing agency;

- loyalty, measured as standard deviation of transactions per potential partner.

All statistical analyses were performed with GenStat 12th Edition (VSN International Ltd., Hemel Hempstead, Hertfortshire). Sensitivity analyses was performed with USAGE 2.0, a collection of GenStat algorithms for sensitivity and uncertainty analysis [1].

4 Results

This section presents results of the sensitivity analysis of simulations with the multi-agent model. All simulations were run with a population of 8 supplier agents and 8 customer agents. The agents were free to select or refuse a trade partner, negotiate and quit negotiations or accept an offer, and deliver truthfully or defect. The simulations ran for 100 time steps. The maximum number of transactions that can practically occur in such a run is between 160 and 180.

For the first series of simulations, parameters sets are drawn at random for configuration of homogeneous agents per run. Cultural indices, relational factors, and the default model parameters referred to in Table 1 are all drawn independently. For each parameter set the model was run 15 times with different random seed, in order to estimate the variance introduced by intrinsic stochastic effects. Statistics are collected at aggregate level. For 627 out of 1000 generated parameter sets the median of the number of transactions equaled zero over 15 replications.

4.1 Probability that Transactions Occur

A logistic regression model [13] was used to investigate which parameters or combination of parameters (interaction) were of significant influence on the probability whether or not transactions occurred (binary data: median equals zero or median greater than zero).

A first exploration revealed that concession factor γ is the most dominant parameter to predict the occurrence of transaction: from 20% for low values of γ to 60% for high values.

Interactions between parameters appeared to play an important role. Starting from a logistic regression model containing all main effects, significant interactions ($p < 0.05$) have been added by forward selection. Table 2 presents the coefficients for the main effects and the significant interactions in the model.

The parameters that have significant effect without interactions are PDI, impatience, and risk avoidance. The probabilities that transactions occur are:

- 0.2789 for $PDI = 0.01$; 0.4025 for $PDI = 0.99$;
- 0.3949 for $\iota = 0.01$; 0.2839 for $\iota = 0.99$;
- 0.4075 for $w_r = 0.01$; 0.2716 for $w_r = 0.20$.

Table 2 Coefficients for main effects (left hand side) and interactions (right hand side) in the logistic regression model of the probability that transactions occur in a simulation run

Parameter	Symbol	Coefficient	Interaction	Coefficient
Power distance	PDI^*	0.566	$\lvert s_a - s_b \rvert \cdot \gamma$	-4.39
Uncertainty avoidance	UAI^*	-0.122	$\bar{s} \cdot \gamma$	-4.43
Individualism	IDV^*	2.015	$MAS^* \cdot v$	-2.581
Masculinity	MAS^*	2.300	$LTO^* \cdot \gamma$	3.134
Long-term orientation	LTO^*	3.02	$LTO^* \cdot MAS^*$	-3.108
Group Distance	D	0.211	$\bar{s} \cdot t_0$	-4.37
Mean status	$\bar{s}$	7.29	$\bar{s} \cdot LTO^*$	-3.38
Status difference	$\lvert s_a - s_b \rvert$	0.762	$\lvert s_a - s_b \rvert \cdot LTO^*$	2.69
Loyalty	B	0.276	$LTO^* \cdot w_q$	-14.18
Learning	C	-0.454	$IDV^* \cdot w_q$	-12.13
Initial preference	J_0	-1.948	$D \cdot LTO^*$	-2.426
Concession factor	γ	4.42	$\lvert s_a - s_b \rvert \cdot J_0$	2.99
Negotiation speed	v	1.015	$J_0 \cdot w_q$	9.59
Impatience	ι	-0.509	$IDV^* \cdot h$	-1.884
Quality preference	w_q	3.27	$e.f$	-2.934
Risk aversion	w_r	-3.22		
Minimal honesty	h	0.296		
Honesty decay factor	f	1.265		
Initial trust	t_0	2.919		
Negative update factor	u_-	-0.062		
Endowment factor	e	1.231		

For parameters that have significant interactions, probabilities can only be predicted if the interactions are taken into account. For instance, the effect of MAS can be predicted in interaction with LTO and negotiation speed v. Table 3 shows that the effect of MAS is great if LTO and negotiation speed are both high or both low.

Table 3 Prediction of the probability that transactions occur with different values of MAS in interaction with LTO and negotiation speed v

LTO^*	MAS^*	$v = 0.01$	$v = 0.99$
0.01	0.01	0.1889	0.3805
	0.99	0.6774	0.3170
0.99	0.01	0.4130	0.6497
	0.99	0.2427	0.0662

For parameters that have significant interactions, probabilities can only be predicted if the interactions are taken into account. For instance, the effect of MAS can be predicted in interaction with LTO and negotiation speed v. Table 3 shows that the effect of MAS is great if LTO and negotiation speed are both high or both low.

From study of interaction tables like Table 3, it is concluded that transactions are unlikely to occur ($p < 0.20$ for extreme values of the parameters) if

- group distance and LTO are both high
- status difference and concession factor are both low
- MAS, LTO, and negotiation speed are all high or all low
- status difference and initial trust are both low
- IDV, LTO, and quality preference are all high
- status difference is low and initial partner preference is high
- initial partner preference is low and quality preference is high
- IDV and minimal honesty are both high
- honesty decay factor and endowment factor are both high

4.2 Sensitivity Analysis

For analysis of the relations between parameters values and outputs a set of 1000 simulations with at least 16 successful transactions is generated. Parameter sets are randomly drawn and used for homogenous configuration of agents in simulation runs, until 1000 runs have produced at least 16 transactions. For each of the 1000 selected parameter sets 15 replications are run. The replications are used for analysis of the variance between parameter sets versus the variance caused by stochastic effects in the replications (Table 4). The percentages are small. the variation between simulations is dominantly caused by parameter variation.

Table 4 Mean variance in replications as percentage of total variance

observed output	% variance	observed output	% variance
number of transactions	1.60	number of deceitful transactions	8.63
number of failed negotiations	3.95	number of traces	7.71
average duration of negotiations	5.32	number of fines	13.75
number of high quality transactions	4.68	average loyalty	5.81

The mean values of outputs of 15 simulations per parameter set are used for analysis. As an example we treat the analysis of the number of transactions. Straightforward sensitivity analysis based on a smoothing spline with two degrees of freedom results in 61.3% of the variance accounted for. For a few parameters the difference between the top and bottom marginal variance is substantial. This can only be due to correlations between parameters (caused by the selection process). Correlation coefficients are small, see Table 5. Therefore, correlations are not further analyzed.

Since 39% of the variation is not explained, several other models are tried, including smoothing splines with 5 degrees of freedom (63.1%), polynomial models, models taking second and third level interactions into account, and log transformations on output and on both input and output. The models using log transformations perform worse than models with polynomial and interaction terms (74.1% and 44.4%, respectively).

Table 5 Parameters having correlation coefficients of 0.10 or more

Parameter	Parameter	Correlation coefficient
Concession factor	Group distance	0.11
	Mean status	-0.14
	Quality preference	0.13
	IDV	0.11
	LTO	0.15
Mean status	LTO	-0.11
Negotiation speed	Quality preference	0.16

The best fit is obtained with a model including quadratic terms and 33 two and three factor interaction terms, that explained 80.7% of the variation. For efficiency the sensitivity analysis is performed with a model with quadratic terms and two factor interactions that explains 79.5%. All parameter combinations in the three factor interactions are also represented as two-factor interactions in the latter model.

For all variables and their interactions both linear and quadratic terms are taken into account (comparable to a smoothing spline with df=2). The result is a model with 30 two-factor, forwardly selected, interaction terms explaining 79.5% of the variation. The interest is not in the model but in its use for gaining insight in the sensitivity of the multi-agent model. Based on this model the bottom marginal variance is calculated for each parameter by leaving this variable and all interaction-terms involving this variable, out of the model. Table 6 presents top and bottom marginal variances.

Table 6 Top Marginal Variance and Bottom Marginal variance of parameters as percentage of the total variance of the number of transactions

Parameter	TMV(%)	BMV(%)	Parameter	TMV(%)	BMV(%)
Index of culture			Loyalty parameter	0.0	0.0
– PDI	0.0	0.6	Loyalty decay factor	0.0	0.1
– IDV	0.2	5.3	Concession factor	9.1	25.0
– UAI	0.8	3.4	Negotiation speed	31.8	39.3
– LTO	0.7	6.8	Impatience	0.6	2.5
– MAS	2.0	7.7	Quality preference	1.5	0.7
Group distance	2.7	6.8	Risk avoidance	0.0	3.0
Mean status	0.3	5.1	Negative update factor	0.9	0.3
Status difference	0.0	1.9	Endowment factor	0.1	0.0
Initial trust	2.7	6.3	Minimal honesty	0.0	0.0
Initial partner preference	1.2	3.0	Honesty decay factor	0.0	0.0

Variation in the number of transactions is for 32% due to variation in negotiation speed. Some other input variables interact with negotiation speed, resulting in a bottom marginal variance of 39%. This means that without good information about negotiation speed 39% of the variation in the number of transactions will remain.

The differences between TMV and BMV of culture and relational factors indicate that these parameters largely have their effect in interactions.

The model developed for the number of transactions cannot be applied for sensitivity analysis of the other output variables. The percentage of variation explained is unsatisfactory. Sensitivity analyses for other outputs have been carried out straightforwardly using smoothing splines (df=2), also resulting in unsatisfactory explanation of variation. The results indicate that modeling steps as applied for the number of transactions need to be followed for each output individually. The parameters contributing at least 10% to output variation according to the sensitivity analysis with smoothing splines (df=2) are given below for each output variable.

- negotiation failure: negotiation speed, concession factor, impatience
- negotiation duration: negotiation speed, UAI, MAS
- quality: initial trust, LTO, quality preference
- deceit: initial trust, MAS
- tracing: MAS
- fines: MAS
- loyalty: initial partner preference, negotiation speed

4.3 Differences between Cultures

As found in the preceeding subsections there are many interactions between parameters. To further analyse the interaction with culture, sensitivity analyses are performed for 62 actual national cultures. 1000 simulations are run for each culture, each simulation with a randomly drawn parameter set that is used to configure a homogeneous agent population. Sensitivity analysis is performed on the simulations that result in at least one transaction. The purpose of this step is to focus sensitivity analysis on the parameters in Table 1 and relational characteristics, and to find differences in sensitivity between cultures.

The number of simulations resulting in one or more transactions ranges from 228 through 490 across cultures, with mean 403. Taking only the runs with a positive number of transactions into account, three different models were fitted per culture (the minimum and maximum percentage of variation explained across cultures is given in parentheses):

- with all 16 parameters linear in the model (minimum 70.7%, maximum 81.2%),
- with all 16 parameters as a spline with 3 degrees of freedom in the model (minimum 73.3%, maximum 83.4%),
- with all 16 parameters and their first order interactions in the model (minimum 83.4%, maximum 89.8%).

No strong nonlinear effects seem to occur in the analyses per culture. Interactions between parameters are present. Table 7 presents some results from the analyses per culture. The sensitivity for parameter changes varies widely across cultures.

Table 7 Mean Top Marginal Variance values (of 62 countries) and data for the countries that have the maximum TMV score for a parameter

national culture	group distance	mean status	initial trust	partner pref.	conces. factor	negot. speed	quality pref.	risk avoid.
Mean (n=62)	4.1	1.0	3.4	1.8	24.8	30.2	0.5	1.6
Indonesia	16.9	0.1	0.2	0.0	11.6	40.9	0.0	0.0
Morocco	0.7	8.7	3.3	4.7	17.5	37.9	0.0	0.0
Hungary	0.0	0.0	11.5	1.9	37.6	1.4	2.4	11.5
Uruguay	4.9	1.7	8.5	5.5	23.4	24.1	0.0	0.0
Netherlands	0.7	0.0	1.6	0.3	46.2	28.8	0.0	2.1
Iran	1.3	3.1	0.9	0.4	10.3	56.2	0.6	0.0
Austria	0.0	0.0	3.8	2.6	27.1	11.5	4.8	6.9
Japan	0.5	0.0	8.3	1.6	30.8	0.9	0.1	15.8

4.4 Aggregate and Individual Level

To perform sensitivity analysis at agent level in heterogeneous agent populations, parameter sets for 4000 simulations are drawn. First 4000 sets of cultural indices, group distance, and status data are drawn. For each simulation run, all agents are configured with equal culture. This restricts partner selection to partners with equal cultural background. For each agent the other parameters are randomly drawn, resulting in a sample of 32000 suppliers and 32000 customers.

A sensitivity analysis could not be completed. Too low levels of explained variation were obtained: for the number of transaction the explained variation was for supplier agents 48.5% with linear fit and with 51.3% smoothing splines; for customer agents 37.6% with linear fit and 38.6% with smoothing splines. However, an interesting result was obtained. The pattern of marginal variance of suppliers matches the pattern found at aggregate level, with negotiation speed and concession factor as dominant parameters. The pattern of marginal variance of customers is very different, with relational characteristics explaining most of the variation. The information asymmetry explains this difference: trust is relevant only for customers.

5 Conclusion

Through sensitivity analysis insight can be gained into the properties of a model.

For exploration of the regions where realistic behavior occurs, logistic regression can be used and probabilities of realistic behavior can be predicted with the model.

Parameters that have significant effects can be identified through metamodeling, even for complex systems. However, the analysis is not straightforward.

The interactions between culture and other parameters are the main cause of the model's complexity. When keeping culture constant, straightforward methods for sensitivity analysis can be applied. Results differ considerably across cultures.

Sensitivity of individual agents can differ considerably from aggregate level sensitivity. However, a method for individual agents has to be developed.

References

1. Goedhart PW Thissen JTNM (2009) Biometris Procedure Library for Genstat 12th edition. Biometris, Wageningen http://www.biometris.wur.nl/UK/Software/Genstat+procedures
2. Hastie TJ, Tibshirani RJ (1990) Generalised Additive Models. Chapman & Hall, London
3. Henning PA (2008) Computational Evolution. In: Schredelseker K, Hauser F (eds) Complexity and Artificial Markets, Lecture Notes in Economics and Mathematical Systems 614, pp 175–193. Springer, Heidelberg
4. Hofstede G (2001) Culture's Consequences, 2nd Ed. Sage Publications, Thousand Oaks
5. Hofstede GJ, Jonker CM, Verwaart T (2009) Simulation of Effects of Culture on Trade Partner Selection. In: Hernández C et al. (eds) Artificial Economics: The Generative Method in Economics, Lecture Notes in Economics and Mathematical Systems 631, pp 257–268. Springer, Heidelberg
6. Hofstede GJ, Jonker CM, Verwaart T (2009) A Multi-agent Model of Deceit and Trust in Intercultural Trade. In: Nguyen NT et al. (eds) ICCCI 2009, Lecture Notes in Computer Science 5796, pp 205–16. Springer, Heidelberg
7. Hofstede GJ, Jonker CM, Verwaart T (2010) Cultural Differentiation of Negotiating Agents. Group Decis Negot, doi: 10.1007/s10726-010-9190-x
8. Jansen MJW, Rossing WAH, Daamen RA (1994) Monte Carlo estimation of uncertainty contributions from several independent multivariate sources. In: Grasman J, van Straten G (eds) Predictability and Nonlinear Modelling in Natural Sciences and Economics, pp 334–343. Kluwer, Dordrecht
9. Jonker CM, Meijer S, Tykhonov D, Verwaart T (2006) Multi-agent Model of Trust in a Human Game. In: Mathieu P et al. (eds) Artificial Economics. Lecture Notes in Economics and Mathematical Systems 564, pp 91–102. Springer, Heidelberg
10. Jonker CM, Treur J (2001) An agent architecture for multi-attribute negotiation. In: Nebel B (ed) Proceedings of the seventeenth international joint conference on AI, IJCAI'01, pp 1195–1201.: Morgan Kaufman, San Francisco
11. Kirman A (2008) Artificial Markets: Rationality and Organisation. In: Schredelseker K, Hauser F (eds) Complexity and Artificial Markets, Lecture Notes in Economics and Mathematical Systems 614, pp 195–234. Springer, Heidelberg
12. Llacay B, Pfeffer G (2009) Foundations for a Framework for Multiagent-Based Simulation of Macrohistorical Episodes in Financial Markets. In: Hernández C et al. (eds) Artificial Economics: The Generative Method in Economics, Lecture Notes in Economics and Mathematical Systems 631, pp. 257–268. Springer, Heidelberg
13. McCullagh P Nelder JA (1989) Generalized Linear Models, 2nd ed. Chapman & Hall, London
14. Ott RL, Longnecker, M (2010) An introduction to statistical methods and data analysis, 6th ed. Cengage, Florence
15. Richiardi M, Leombruni R, Saam NJ, Sonnessa M (2006) A Common Protocol for Agent-Based Social Simulation. Journal of Artificial Societies and Social Simulation 9(1):15 http://jasss.soc.surrey.ac.uk/9/1/15.html
16. Saltelli A, Chan K, Scott EM (2000) Sensitivity analysis. Wiley, Chichester
17. Saltelli A, Tarantola S, Campolongo F, Ratto M (2004) Sensitivity analysis in practice. Wiley, Chichester
18. Weisbuch G, Kirman A, Herreiner D (2000) Market Organisation and Trading Relationships. Economic Journal 110:411–436
19. Williamson, OE (1998) Transaction Cost Economics: how it works, where it is headed. De Economist 146:23–58

LaVergne, TN USA
09 September 2010
196514LV00002B/21/P

9 783642 139468